Prentice Hall's Exploring Biology Series

GENES, AGING, AND IMMORTALITY

W9-DDQ-662

CHARLOTTE A. SPENCER

UNIVERSITY OF ALBERTA

PEARSON

Prentice Hall

Upper Saddle River, NJ 07458

Editor-in-Chief, Life and Geosciences: Sheri L. Snavely
Executive Editor: Gary Carlson
Project Manager: Crissy Dudonis
Editorial Assistant: Jennifer Hart
Executive Managing Editor: Kathleen Schiaparelli
Assistant Managing Editor: Becca Richter
Production Editor: Rhonda Aversa
Supplement Cover Manager: Paul Gourhan
Supplement Cover Designer: Joanne Alexandris
Manufacturing Buyer: Ilene Kahn
Director, Image Resource Center: Melinda Reo
Manager, Rights and Permissions: Zina Arabia
Image Specialist: Beth Boyd-Brenzel
Image Permission Coordinator: Joanne Dippel
Cover Photograph: Colin Anderson/Brand X Pictures/Getty Images

© 2006 Pearson Education, Inc.
Pearson Prentice Hall
Pearson Education, Inc.
Upper Saddle River, NJ 07458

Printed in the United States of America

ISBN   0-13-187413-6

Pearson Education Ltd., *London*
Pearson Education Australia Pty. Ltd., *Sydney*
Pearson Education Singapore, Pte. Ltd.
Pearson Education North Asia Ltd., *Hong Kong*
Pearson Education Canada, Inc., *Toronto*
Pearson Educación de Mexico, S.A. de C.V.
Pearson Education—Japan, *Tokyo*
Pearson Education Malaysia, Pte. Ltd.

Table of Contents

Preface

The desire to conquer aging and extend the human lifespan is as ancient as recorded history. As far back as 3500 BC, purveyors of antiaging elixirs promised youthful appearance and renewed vigor. Both Alexander the Great and Ponce de León risked their lives to seek the legendary Fountain of Youth. Religion and literature accept the notion that immortals may live amongst mortals. Stories, such as those of Faust and Dracula, demonstrate that people will bargain their souls to gain immortality. Even people in our modern age will endure considerable pain and reduced quality of life to prolong their lives through an obviously terminal illness. People spend billions of dollars each year attempting to arrest the aging process with diets, hormones, antioxidants and extreme treatments in longevity clinics. Now that the baby boomers are approaching their senior years, the subject of aging is increasingly becoming a focus of public discussion.

Until very recently, the study of aging was tarnished by its historical association with charlatans and "snake oil" remedies. Only within the last decade have scientists made significant progress on dissecting the causes of aging. By manipulating the genetic makeup of laboratory organisms such as yeast, fruit flies and mice, scientists have discovered that some genes, when mutated, can extend the lifespans of these organisms up to five-fold beyond normal. A simple dietary manipulation – calorie restriction – will also increase longevity, at least in experimental animals. These approaches are pinpointing the basic physiological causes of the aging process, and may soon be translated into interventions that may increase the maximum lifespan of humans.

In this booklet, we will describe some of the current progress in the scientific study of aging. We will answer questions about aging's basic biology, as well as the prospects for human immortality. Social and ethical questions surrounding the real possibility of human lifespan extension will be explored. We will also relate some fascinating stories that surround the very human desire to conquer aging and gain immortality.

The information in this booklet is as current and accurate as possible, and is derived from scientific literature, media and government sources. The study of aging is advancing extremely rapidly; therefore, readers are encouraged to refer to the publication and internet sources listed in the References and Resources section for the latest developments on genes, aging and immortality.

Introduction

From the moment of birth, time delivers us closer and closer to death. As we move from youth to middle and old age, our bodies and minds gradually weaken, and ultimately cease to function. To make the process worse, we humans are painfully aware of our own progressive decline and inevitable death. But the drive to live is strong, and we grasp for anything – myths, legends, magic potions and now technology – that will end disease and suffering, extend our lives and cheat death (Figure 1).

The desire for youth and longevity is ancient. Like many people today, the Sumerian god-king Gilgamesh longed for immortality, but even he could not attain it. After picking the magical plant of rejuvenation, Gilgamesh paused to relax in a pool of water, leaving the plant on the shore. To his dismay, a passing serpent devoured his elixir of life. Ancient Taoists consumed foods such as eggs, peaches and tree parts which they believed would combat the aging process. They rigidly withheld their breathing, in the belief that controlled breath would nourish the body in the absence of food and lead to immortality.

Roger Bacon, a 13th century philosopher and scientist, believed that the symptoms of old age were caused by a loss of internal moisture, and that longevity could be enhanced by replacing this moisture – found in substances such as pearls, coral, aloe wood and bones from stags' hearts. He also encouraged older men to spend time in the presence of young virgins, so that they could absorb virgin's breath – thought to be an important source of vital energy. The alchemists also believed that immortality could be achieved and that the Fountain of Youth was a reality. In their efforts to achieve a longer life, they promoted the ingestion of antiaging substances as precious as gold and as toxic as cinnabar (mercuric sulfide). The ultimate goal of the alchemists was to create a substance with even more powerful antiaging properties than gold. Although they believed that they could create this substance by applying the mythical Philosopher's Stone, both the Stone and immortality eluded them.

More recent quests for elixirs of longevity seem as bizarre, and as futile, as those of the past. In the late 19th century, a French physiologist named Charles E. Brown-Séquard offered his elderly patients a treatment that he claimed would reverse the symptoms of aging. His elixir consisted of extracts prepared from the crushed testicles of young dogs, rabbits and guinea pigs. After injecting these extracts into himself and several other elderly men, he reported that he had regained his youthful strength and mental acuity, as well as improved control over his bowels. His other patients claimed similar improvements in mental and physical performance. Brown-Séquard's experiments led to the rise of "organotherapy" in Europe and North America. Extracts from various animal tissues – from spleen to pituitary – were injected into patients to treat diseases and reverse aging. Recent analyses of Brown-Séquard's experiments reveal that the benefits of organotherapies were entirely due to placebo effects.

Interestingly, testicular treatments remained a popular remedy for aging well into the 20th century. American self-proclaimed physician, snake-oil vendor and gubernatorial candidate John Brinkley performed thousands of transplants of goat testicles into older men. Although his

Figure 1

In 1512, the Spanish explorer, Ponce de León, traveled around the coast of Florida searching for the Fountain of Youth. Although he and his crew drank water from numerous lakes and springs, the elixir of longevity eluded him. Source: Bettmann/Corbis

patients claimed to have gained greater vitality, they also complained that they had acquired a strong desire to "chew sprouts." The State of Kansas revoked Brinkley's medical license in 1930.

In the 1920s, European physiologist Eugen Steinach enhanced the lives of his patients by performing vasectomies and by grafting testicles of young men onto older men – treatments thought to increase the levels of youth-enhancing testosterone. Other popular European vitality treatments involved applying electric currents and radioactivity to male sex organs.

Many of today's popular longevity treatments bear an eerie resemblance to those of the past. Hormones, antioxidants, wrinkle creams, dietary supplements, injection of fetal sheep cells, breathing hot radioactive air in mine shafts are all believed to slow or reverse aging. And in case these methods are ineffective, some believe that cryonic freezing of those declared dead will lead to immortality – as the frozen await a future time when death, and freezing, can be cured and the patient can be returned to life.

Although dietary supplements, youth-bestowing waters and magic elixirs predominate in the search for a long life, specific behaviors and lifestyles are a close second. Lives of simplicity, restraint and righteousness are claimed to be a prerequisite for a long life. Physical exercise, moderation in diet and low stress levels are also thought to contribute to longer, healthier lives.

However, despite our attempts to combat aging and prevent death, we remain powerless to stop them. Aging and death are deeply mysterious, complex biological processes that are only beginning to reveal their secrets. Although science is making progress in aging research, a full understanding of why we age and die remains uncertain.

But, is there hope? Must we grow old? Can modern science help us understand the why's and how's of aging and death? Can new technologies, one day, help us achieve immortality?

In this booklet, we answer questions about the mysteries of aging and the search for immortality. The answers to these questions suggest that, in the near future, genetics and biotechnology may translate the scientific understanding of aging into treatments that will significantly extend the human lifespan. We will consider how we will deal with these longer, and hopefully healthier, lives and how these longer lives will change social and cultural attitudes. After millennia of searching, we may be at the beginning of a genuine longevity revolution.

Questions About the Biology of Aging

WHAT IS AGING?

Aging is defined as a process of general, irreversible and progressive physical deterioration that occurs over time. This deterioration leads to increased susceptibility to disease and probability of death. It is a process that manifests itself in every part of the body, in individual cells, and in molecules such as DNA, lipids and proteins.

While trying to define aging, it is important to differentiate between the symptoms of aging and the symptoms of diseases that accompany aging. For example, older people generally have slower mental processes and weaker short-term memories. However, these features are not restricted to the elderly, and many older people do not develop significant slowing of mental capabilities. The probability of contracting diseases such as Alzheimer's disease, arthritis, cancer, heart disease, and osteoporosis increases as people age; however, the symptoms of these diseases are not the same as the symptoms of aging itself.

Although we think that we intuitively recognize the symptoms of aging, the process is multifaceted and varied. Most of the physical and mental changes that accompany aging are not uniform in timing or in intensity between individuals. Some people may develop gray hair or wrinkled skin in their twenties or thirties, whereas some people in their eighties still have colored hair and smooth skin. Some people in their nineties have keen mental abilities, whereas others in their fifties may have difficulties with memory or responsiveness.

Measuring Aging, Lifespan and Life Expectancy

The most widely utilized measure of aging rate, in both individuals and populations, is age-at-death. Although this characteristic is easy to measure, it does not directly measure the complex symptoms of physical decline that may occur prior to death. However, research has shown that age-at-death, at least in controlled experimental situations or in industrialized societies, is a reasonable measure of aging rate.

In populations, aging is often represented by survival curves. These curves show the percentage of individuals that survive as a function of their age. There are two types of theoretical survival curves (Figure 2). One type (curve A) represents a population in which all deaths are due to the effects of aging. In this population, most individuals survive past reproductive age and then begin to die over a short period of time. The other type (curve B) represents a population in which all deaths are due to accidental causes and aging has no effect. Accidents, predation and infectious diseases remove a constant proportion of the population each year. This population declines exponentially, so the survival curve would appear as a straight line on a logarithmic scale. Survival curve B is the type of survival curve seen for animals in the wild that rarely survive long enough to experience the effects of aging.

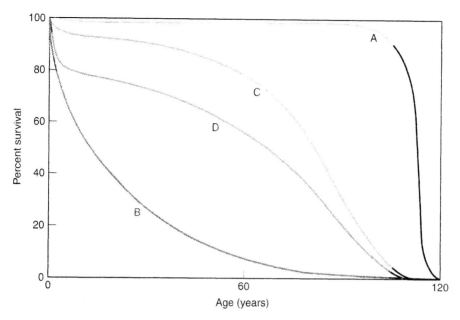

Figure 2
Survival curves for theoretical (A, B) and actual (C, D) human populations.

Real survival curves for human populations show combinations of accidental and age-related deaths. Survival curve C is the current survival curve for persons born in the United States. It closely resembles curve A more than curve B, as most people in western industrialized countries die of age-related causes. Curve C shows an initial rapid loss of individuals, due to infant mortality. After infancy, few deaths occur until aging begins – perhaps beginning as early as the mid-twenties. In addition, there is a slight flattening towards the end of the curve. Apparently, people who live to advanced ages have a higher than average probability of surviving to even greater ages. These longest-lived individuals may be genetically resistant to some or all of the effects of aging, or have a genetic aging program that is relatively slow. In support of this idea, humans who survive beyond the age of 85 have a ten-fold lower rate of heart disease and cancer than people between the ages of 65 and 85.

Curve D represents the survival curve of people born in the United States in the 19th century. This curve more closely resembles curve B, with accidental death being a significant cause of mortality at all ages. Prior to the 20th century, there were much higher rates of infant mortality, infectious diseases and accidental deaths than those seen today.

Interestingly, the maximum age in both real human populations (curves C and D) is about the same – 100 to 120 years. Maximum lifespan in humans is the age of the oldest individual. Currently, the oldest person whose age has been verified is Jeanne Calment who died in 1997 at the age of 122 (See Box: Tale of a Supercentenarian).

Scientists also measure longevity in terms of average lifespan, also referred to as life expectancy. Life expectancy is the average lifespan of all individuals in the population. Life expectancy is strongly influenced by factors such as infant mortality rates, malnutrition, wars and

Box — Tale of a Supercentenarian

There are more centenarians alive now than at any time in history; many of them are healthy, independent and active people. The story of the world's oldest authenticated centenarian illustrates how mental and physical vitality can accompany a life of extreme longevity.

Jeanne Calment was born in Arles, France on February 21, 1875. In the same year, Tolstoy published Anna Karenina and the first telegraph cable joined Britain and the United States. Calment apparently met Vincent Van Gogh in her father's shop in Arles, but said that she "was not impressed." During her lifetime, Jeanne Calment witnessed the invention of the automobile, telephone, air travel and use of electricity.

Longevity ran in Calment's family. Her father died at the age of 94 and her mother died at age 86. Calment married her cousin, who was a successful store owner. He died of food poisoning in 1942 at the age of 46, leaving her a widow for more than 50 years. Her only child, a daughter, died of pneumonia in 1934. Calment raised her grandson and he died in 1963 in an automobile accident.

Calment never worked during her life. She came from a prosperous family and spent her time in pursuits such as tennis, swimming, piano playing and the arts. During the last 30 years of her life, she lived on an income provided by a local lawyer. In 1965, André-François Raffray entered a property deal that he would later regret. In that year, Calment was 90 years old and living in a house in Arles. Raffray, who relied on the predictions of actuarial tables, agreed to pay Calment $500 a month until she died, in exchange for inheriting her house. Raffray died in 1996 at the age of 77. His family continued to pay Calment for more than a year until she died. In all, Raffray and his family paid Calment about $180,000 – three times the value of the house. On her 120th birthday, Calment teased Raffray by telling him, "We all make bad deals in life."

It is not clear what accounts for Calment's extreme longevity. Calment herself felt that her indifference to stress, her occasional glass of port wine and her diet rich in olive oil were responsible. However, she also ate about two pounds of chocolate each week and smoked until the age of 119. She would have continued smoking, but she was going blind and was too proud to ask someone else to light her cigarettes. She rode her bicycle until the age of 100 and took up fencing at 85. Although going blind and losing her hearing in her last few years, Calment remained mentally alert until her death. At the age of 121, she released two CDs documenting her life and thoughts. Many of her quotations indicate her sense of humor and joy in life:

- On her 120th birthday, she was asked what kind of future she expected, to which she replied, "A very short one."
- She also said, "I've waited 110 years to be famous. I count on taking advantage of it."

Jeanne Calment, at age 121, showing off her new CD. She died in 1997 at the age of 122 years, 164 days. Source: JPP/STR Reuters environmental disasters. In situations where accidental deaths are commonplace, life expectancy values do not reflect aging rates.

Aging Has Become a Significant Feature of Human Life Only in Recent History

In the wild, most animals do not survive long enough to experience significant aging. Disease, malnutrition and predation terminate the lives of most animals before they reach old age. Only those animals living in the protected environments of homes, zoos and laboratories experience the symptoms of aging.

Aging is a relatively new experience for humans as well. As recently as 10,000 years ago, most people did not live beyond the age of 30. Even by the 17th century, Europeans had a short life expectancy. A study performed in 1693 revealed that only 51 percent of newborns survived until the age of 10, 28 percent survived into their fifties and 11 percent survived to the age of 70. Even in the early 1900s, about half of all children died before the age of 14. The life expectancy in the United States in 1900 was about 45 years. Over the last century, there has been a significant increase in life expectancy – developed countries experienced an increase of about 30 years from 1820 to 1980. At present, in the developed world, over 90 percent of babies survive to at least age 50. In contrast, people in developing nations have a life expectancy of approximately 15 to 20 years less than those in developed countries. This is mostly due to high infant mortality as a result of infectious diseases such as malaria, tuberculosis, influenza and measles.

In developed countries, current life expectancies lie between 78 and 85 years for women and between 73 and 80 years for men. Japanese women have the world's longest life expectancy at 85.2 years. Japanese men are also high in the rankings for male longevity, at 78.2 years. The low levels of cancer and strokes, perhaps due to traditional low-fat diets, may contribute to the long lives of Japanese people. Other record-holders are women in Hong Kong (84.6 years), France (82.5), Australia (82.4), Italy (82.1) and Canada (82.0). Men in Hong Kong (78.4 years), Japan (78.2), Iceland (78.1), Sweden (77.73), Australia (77.0) and Switzerland (76.9) round out the list of longest-lived people in the world. In the United States, the life expectancy for Caucasian males is 74.9 years and for Caucasian females is 80.1 years. The life expectancies for all races in the United States are 74.4 for males and 79.8 for females. The World Health Organization states that several factors contribute to the relatively low US life expectancy. These factors include the poor health of Native Americans, rural African Americans and inner city poor, as well as high rates of tobacco-induced cancers, heart disease and violent homicides.

At the bottom of the life expectancy scale are people in sub-Saharan Africa. Due to AIDS, malaria, tuberculosis and other infectious diseases, people in these regions have a life expectancy of between 44 and 46 years.

At present, there is an approximately five to seven year gap between the life expectancies of men and women in developed countries. In Russia, the gap is even greater – about ten years. Scientists think that part of the gender difference may be genetic, as males have higher death rates at all ages, even prior to birth. Higher male death rates may also be due to behavioral differences, as young males tend to partake in high risk activities. Other reasons may be that women are more health conscious, consume better diets and smoke less than men. The extreme difference between male and female life expectancies in Russia is thought to be due to the high levels of male alcohol abuse, which leads to high rates of heart disease, violence and accidents.

WHAT HAPPENS TO OUR BODIES AND CELLS AS WE AGE?

Our Aging Bodies

As we age, we become increasingly aware of changes occurring in our bodies and minds. We perceive most of aging's manifestations as deleterious. Many changes, such as deterioration of arteries, brain cells and kidney function, are serious and have the potential to be life-threatening. However, some changes such as gray hair, incontinence or wrinkles may have only social or psychological effects.

Among the visible signs of aging are wrinkled skin, gray hair, loss of strength and a decrease in size. Skin wrinkling is due to the thinning of the skin epidermis and reduced elasticity of skin. Skin becomes thinner because of reduced levels of subcutaneous fat. Elasticity declines because of alterations to the levels and structures of collagen and elastin proteins. Also, skin becomes dry because oil glands become less effective. Hair becomes gray or white because melanocytes in hair follicles become impaired, leading to loss of hair color. Beginning at about age 40, muscles weaken by one to two percent per year and this muscle is often replaced by fat. Muscle mass declines and strength may be reduced up to 40 percent by the age of 80. This loss of strength may lead to falls and bone fractures. Urinary incontinence affects millions of older people and may lead to embarrassment and social isolation. Older people often appear smaller, or wizened, as a result of bone loss. Bone mass is lost in many older people at a rate of 1 to 2 percent annually. Osteoporosis contributes to about 20 percent of fatal hip fractures in the elderly.

Almost every physiological system in the body experiences age-related changes. Lung capacity declines. Oxygen transfer is impaired due to loss of flexibility of collagen and elastin proteins within the lung. The elderly are less sensitive to insulin, growth factors and other hormones. The levels of many hormones such as sex hormones and growth hormone decline with age. Hearing, taste and smell become less sensitive. More than 70 percent of people over 65 require corrective lenses because the eye's lens becomes thicker and more rigid, leading to difficulties seeing close objects. Muscles in the iris weaken and pupil size is reduced, resulting in the requirement for more light. Sleep patterns change. Most older people have shorter sleep periods, more waking periods during the night, and less REM (or dream) sleep. Kidney function declines with age. Kidney weight decreases about 15 percent between the ages of 40 and 80. Brain function also declines in the elderly. On average, there is a 10 percent loss in brain weight between the ages of 20 and 90. However, not all elderly people suffer dementia and some studies find no decline in brain weight or IQ in those who have no identifiable brain disease. Since cardiovascular disease greatly affects brain function, some effects of age on the brain may be indirectly due to circulatory changes.

The cardiovascular system may suffer life-threatening conditions such as atherosclerosis, or hardening of the arteries. Fatty deposits and cholesterol accumulate in the arteries, reducing blood flow, raising blood pressure and straining the heart. Clots may disengage from blood vessels and obstruct the flow of blood to the brain or other organs, leading to strokes or heart attacks. The immune system becomes less efficient, contributing to increased susceptibility to infections and cancers. As the immune system becomes impaired in its ability to discriminate between self and non-self proteins, more autoimmune conditions such as arthritis appear. Declines in reproductive capacity occur with age as a result of reduced synthesis of reproductive hormones. Women experience a rapid loss of estrogen and progesterone synthesis. This leads to infertility and menopause usually by the early 50's. Men also experience declines in testosterone levels and fertility as they age.

Diseases Associated with Aging

The incidences of many diseases increase with age. Perhaps the most significant age-related condition is cardiovascular disease, which is responsible for about 43 percent of deaths in those between the ages of 65 and 74. Strokes and gangrene also result from degeneration of the circulatory system. Many people over 65 have a degree of insulin resistance or subclinical diabetes. Some degree of dementia, frequently resulting from Alzheimer's disease and strokes, often afflicts elderly people. Arthritis, psoriasis and autoimmune diseases increase with age. Cancer accounts for about 30 percent of deaths at the age of 65, but only 12 percent at age 80, due to the increased rates of heart disease and Alzheimer's disease in this age group. Despite the apparent leveling off of cancer incidence beyond age 80, cancer rates increase exponentially with age in the absence of other diseases.

Biomarkers of Aging

In order to carry out meaningful research into the causes of aging, scientists must be able to precisely identify the biological changes that occur in the body as a result of aging. They need to measure the rates of these changes and why they occur.

Studies on the causes and prevention of aging are particularly difficult, as there are no reliable, uniform features of aging (biomarkers) that can be accurately measured. A good aging biomarker would be a biological trait that is independent of age-related diseases, progresses over time and is associated with increased mortality. None of the symptoms or diseases of aging described above is uniform in timing or intensity between individuals or exclusively due to aging. To date, no biomarkers of aging have been accepted by the scientific community, except for mortality rate (as described in the previous section).

WHAT DOES IT MEAN TO DIE OF OLD AGE?

If humans did not age and the only causes of death were accidents and infectious diseases, in theory, we would have a life expectancy of about 1,200 years and a maximum lifespan of up to 25,000 years. Obviously, aging has a significant effect on mortality. The likelihood of dying as a result of contracting diseases such as Alzheimer's disease, cardiovascular diseases, cancer and diabetes, increases with age. Even "accidental" deaths can be related to aging, as a general physical weakening increases the chance that an organism will suffer trauma or contract an infectious disease. But can age itself, in the absence of detectable disease, cause death? In other words, is it possible to die of old age?

The general deterioration that occurs with age suggests that one or more organ systems would eventually fail, even in the absence of a definable disease. This is supported by clinical observations that some older humans and animals can die suddenly of no single apparent cause. Animals that are maintained on calorie restricted diets often have longer than normal life spans and are relatively free of age-related diseases. When these animals die, autopsies often fail to detect any disease responsible for their deaths. These animals simply appear to die of organ failure.

In humans, as in other animals, the body adapts to external and internal changes so as to maintain overall homeostasis – a state of biological equilibrium. In the elderly, this homeostatic balance can be rapidly disrupted. As a result, an older person may lose control of blood pressure, blood glucose levels, inflammatory responses, or release of stress hormones, leading to a sudden death. As modern medicine controls and cures the current leading causes of death in older people, it seems likely that other degenerative conditions will simply take their place. Even if science and medicine eliminated all age-related diseases, humans would still age and die. The only way to effectively extend the maximum lifespan of humans, and other animals, will be to slow or stop the fundamental causes of aging itself.

WHY DO WE AGE AND DIE? WHAT IS THE PURPOSE?

Aging is not an inevitable consequence of being alive. Single-celled prokaryotes such as bacteria do not age. They simply die from accidents and starvation. Bacteria divide by an asexual process known as fission – the simple replication of their circular DNA genomes and the distribution of these two genomes into two newly divided cells. Since bacteria comprise about half of the living material on earth and occupy a wide range of biological niches, it is clear that aging is not necessary for the development of a successful life form. Also, aging is not a significant problem for most animals in the wild, as their lives are shortened by natural hazards and starvation. Until recent times, most humans did not experience the debilitating effects of aging, since they too experienced dangerous, short lives.

So, why did nature invent this phenomenon? Is there some evolutionary advantage to aging? Is it programmed, or is it a side-effect of nature's indifference to organisms that have passed their reproductive ages?

One hypothesis states that aging is programmed into every species in order to limit population growth and give room for the next generation. In this scenario, aging and death open up environmental niches and resources so that evolution has new material on which to operate, hence facilitating adaptation to new environments. This hypothesis was once widely accepted, but is now rejected because of current knowledge about how evolution works. It is now accepted that evolution operates by favoring genetic traits that enhance individual fitness leading to reproductive success. There is no mechanism by which evolution can select for traits that enhance the lives of organisms that have passed reproductive age. Also, because most animals do not experience aging in the wild, natural selection would have few opportunities to either favor aging or select against it.

In the early 1950s, British immunologist Peter Medawar proposed that aging occurs precisely because of the inability of natural selection to operate on genetic effects that occur after reproductive age. He proposed that many of the genes that maximize an individual's reproductive fitness do not operate to maintain the body past this time. However, nature tolerates these deleterious age-related effects because there is no way for natural selection to modify them. In this scenario, aging is the result of nature's indifference to what happens after an organism reproduces. It would be wasteful to divert energy into longevity mechanisms when natural hazards are high and most organisms die before aging. In this view, there is no evolutionary advantage in protecting older organisms from the deterioration that leads to cardiovascular disease, cancer, diabetes or Alzheimer's Disease.

A variation of this hypothesis – called antagonistic pleiotropy – was proposed by the evolutionary biologist George Williams in 1957. Pleiotropy refers to the fact that a single gene can have several different effects within an organism. Antagonistic pleiotropy states that some genes may enhance reproductive efficiency early in life; however, these same genes may have serious negative side-effects later in life. Because natural selection fixes these genes in the population for the purposes of reproduction, and the late, negative side-effects are not selected against, the negative side-effects persist. Both Medawar and Williams' hypotheses could be summarized as a trade-off of genetic resources. Nature puts its efforts into making reproduction as efficient as possible and does not protect the organism against any deleterious side-effects that appear in later life. British scientist Tom Kirkwood expressed these ideas as a "disposable soma" theory. Because an individual's genes have been successfully passed on to the next generation, the body (soma) can be discarded.

Antagonistic pleiotropy predicts that organisms that experience high rates of predation and accidental death would have short lifespans. In general, this seems to be the case. For example, mice in their natural environments die from predation, starvation, and cold within a few months of birth. Therefore, nature ensures that mice can reproduce quickly – usually within a month of birth. The speed at which they reach reproductive age may have a price, though, as mice live only about 2 years. In contrast, bats are under less pressure from predation due to their ability to fly. Therefore, bats are able to reproduce later in their lives. Perhaps as a consequence, some of their genes can be selected for longer lives – which are about ten-fold longer than those of mice.

More support for the link between lifespan and predation comes from observations of protected wild populations such as the opossums that live on islands off the coast of Georgia. These island populations, having been free of predators for generations, have longer lifespans and slower aging rates than their mainland cousins.

An interesting study of fruit fly (*Drosophila*) lifespan also supports the proposed relationship between age of reproduction and maximum lifespan. These studies, carried out by Michael Rose at the University of California, Irvine, tested the hypothesis that lifespan could be extended by delaying reproduction. He bred *Drosophila* over many generations by selecting eggs laid by flies that were near the end of their reproductive periods. By gradually increasing the age of egg collection, he eventually bred a strain of flies that preferred to mate later in life. These late-reproducers had low fertility early in life and higher fertility later. These flies lived about 30 percent longer than their unselected relatives (43 days vs. 33 days). Unfortunately, the fitness of these flies suffered from the genetic changes that led to their longer lives. When returned to the wild (a garbage pile with other flies), these slow-breeders died out.

HOW DO WE AGE AND DIE?

There are no shortages of ideas to explain how we age. One survey catalogued over 300 aging hypotheses. Given the large number of complicated biological changes that occur in aging cells and in multicellular organisms, it is quite likely that many of these hypotheses have merit and are not mutually exclusive. Most of the currently favored ideas explaining the mechanisms of aging can be described as variations of either the ***cellular senescence hypothesis*** or the ***wear and tear hypothesis***. In order to understand these two types of hypotheses, it is first necessary to understand how cells die.

Cell Death

Ultimately, aging in all organisms, including humans, is the result of aging of individual cells. An adult human is composed of about 10^{14} (one hundred trillion) cells, each one a part of a tissue type with specific structural features and functions. Liver cells, neurons, and immune system stem cells all have their own unique repertoire of biochemical functions that contribute to the life of the larger multicellular organism.

There are two ways in which cells die. One type of cell death is accidental death. If a cell is deprived of nutrients or oxygen, or is physically damaged, it dies by a process known as necrosis. During necrosis, a cell releases distress signals, then gradually breaks apart. In multicellular organisms such as humans, the release of cellular contents can trigger inflammatory reactions that may lead to tissue damage. If a region of necrosis is large enough, the body lays down scar tissue to replace the dead cells.

The second type of cell death is programmed cell death, also known as apoptosis. Apoptosis is a kind of cellular suicide and is genetically controlled. The signals that tell a cell to commit suicide are quite varied. During early embryogenesis, apoptosis is an essential process that is necessary for proper organ development. An example is the development of the nervous system. In order to establish nerve fiber connections during fetal development, neurons in the brain and spinal cord send out nerve fibers in all directions. The fibers that contact cells in appropriate peripheral tissues are maintained throughout life. However, neurons that do not make appropriate connections are eliminated by apoptosis. Apoptosis also contributes to the operation of the immune system, which manufactures white blood cells (called lymphocytes) in vast numbers. Apoptosis performs essential functions in the immune system by eliminating lymphocytes that have the potential to react with self-antigens (which would lead to autoimmune diseases) and lymphocytes that are no longer required for an immune response to infections or foreign antigens. In addition, lymphocytes that are unable to interact with foreign antigens are eliminated from the system by apoptosis. In both the developing nervous system and the immune system, the instructions to commit suicide are genetically programmed and are essential for the normal functioning of these systems.

Programmed cell death also protects multicellular organisms from cancer and infection. Apoptosis is triggered when DNA is damaged, when cells begin to develop into cancer cells, or when cells are infected with viruses. Specific cellular proteins (described in the next section) sense damage to cellular components and send signals down elaborate biochemical pathways leading to apoptosis of the damaged cells. These mechanisms, and the genes that control them, are highly conserved in evolution – from *Paramecia* to humans.

Cells that die by apoptosis act quite differently than those that die by necrosis. The first visible sign that a cell is committed to suicidal death is that its nuclear DNA breaks into thousands of tiny pieces. At this point, the cell may still be involved in basic metabolic functions such as energy generation and protein synthesis, but it is essentially "brain dead." Soon, the nucleus fragments, followed by the entire cell, which breaks into discrete, small pieces called apoptotic bodies

Figure 3
A normal white blood cell (below) and a white blood cell undergoing programmed cell death (apoptosis). Apoptotic bodies appear as grape-like clusters on the surface of the cell. Source: Science Photo Library

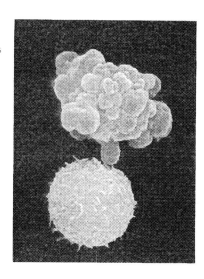

(Figure 3). The surrounding cells engulf the apoptotic cell, and the dead cell's internal contents do not spill into the surrounding tissue. The apoptotic cell simply disappears.

Cellular Senescence Hypothesis

A fundamental question is whether aging is the result of inherent mechanisms, internal to the cell, or whether it results from external assaults from the environment.

The first observations that suggested an inherent biological basis for aging were made over 40 years ago by Leonard Hayflick and his colleagues. These investigators were attempting to establish cell cultures from human primary tissues. They released cells from tissues, placed the cells into sterile culture dishes and allowed the cells to grow in the presence of rich sterile growth media. Cells derived from embryonic tissues grew and divided well for a time, but then ceased to grow. After about 40 to 60 population doublings, the cells became enlarged, flattened and could not be stimulated to divide. If the cells were frozen after a few doublings and thawed at a later date, they resumed dividing, but then stopped dividing on the same schedule – as if they had not been frozen. They appeared to have an internal cell replication counter that instructed the cell that its replicative life was over after a set number of cell divisions. Cell cultures derived from adult tissue donors underwent fewer doublings – 30 or fewer – than those derived from embryonic tissues. Cells derived from organisms with short lifespans, such as chickens or mice, underwent fewer doublings than cells derived from longer living organisms, such as humans or tortoises.

Cells that reach the end of their replicative life are said to undergo replicative senescence. They can remain alive for long periods of time in culture, but cannot re-enter a growth phase. In addition, they have distinct morphological and metabolic characteristics. Cells derived from all normal tissues undergo replicative senescence in culture. The only cells that escape senescence are those that are "transformed." Cells that are infected with certain viruses or cells that undergo mutations in cancer-causing genes can become transformed and grow indefinitely in culture. In addition, cells derived from various cancers are transformed and do not undergo replicative senescence in culture.

What is the molecular nature of this internal cell replication counter? Over the last 15 years, research has revealed that these replication counters are located at the ends of chromosomes – at structures known as telomeres. Telomeres contain short DNA sequences (TTAGGG in vertebrates) repeated numerous times, amounting to a final length of 5 to 15 kilobases. Telomeres form caps on linear chromosomes and prevent chromosome ends from fusing to other chromosome ends. Chromosome fusions create genomic instability, which in turn can trigger mutations in genes that contribute to the development of cancer. Besides preventing chromosome fusions, telomeres perform another essential service. Because eukaryotic chromosomes are linear, and DNA polymerases are unable to replicate the ends of linear DNA molecules, chromosomes become shorter with each DNA replication. If not corrected, this chromosome shortening would eventually lead to deterioration of the genetic material. In order to retain intact chromosomes with full-length telomeres, germ cells contain the enzyme telomerase, which adds telomere DNA repeats onto the ends of chromosomes, conserving telomere and chromosome length. However, most adult somatic cells do not contain telomerase. As a result, about 50 to 200 base pairs of telomeric DNA is lost every time a somatic cell replicates. Once telomeres shrink to a certain critical length, the cell interprets this as an instruction to enter replicative senescence.

The molecular mechanism by which telomere shortening brings about replicative senescence is a recent discovery that provides an interesting link between cancer and aging. It appears that shortened telomeres trigger the activity of a protein known as p53. p53 is a transcription factor – a protein that regulates when and how often a gene is copied into an RNA molecule, which is the first step in gene expression leading to the synthesis of the protein encoded by the gene. The p53 transcription factor regulates the transcription of over 50 different genes. The p53 protein becomes activated in response to DNA damage induced by radiation, chemicals, oxidative stress or other internal or external agents. In normal cells, p53 triggers one of three different responses: transient arrest of cell division, apoptosis or senescence. Cell division arrest allows the cell sufficient time to repair its damaged DNA before it divides, thereby reducing the number of mutations that could be passed to daughter cells and that might lead to cancer. If DNA is severely damaged, p53 may induce apoptosis, thereby removing the damaged cell from the organism. This also protects the organism from developing cancer because the cell with damaged DNA dies before accumulating mutations in cancer-causing genes. If p53 triggers senescence, the senescent cell may remain alive, but cannot divide and therefore cannot form a tumor. Support for the hypothesis that p53 is a tumor suppressor protein comes from the observation that the majority of cancer cells contain deletions or mutations in the *p53* gene. In addition to contributing to cancer formation, the loss or inactivation of the p53 protein extends the lifespan of human cells in culture. In contrast, senescent cells contain elevated levels of p53 activity. Mice that are engineered to have excessive p53 activity appear to have unusually low cancer rates, but they also show symptoms of premature aging and have a 20 percent shorter lifespan.

To summarize, replicative senescence limits the lifespan of cells in culture. This appears to be due to telomere shortening over a number of cell divisions, with p53 providing the signal that triggers the growth arrest.

Does replicative senescence explain aging? Not all cells in the mature body divide and, therefore, would not be susceptible to replicative senescence. For example, specialized differentiated cells such as neurons, endocrine cells, muscle cells and differentiated cells of the immune system do not undergo many, if any, divisions. However, other tissues require continuous replenishment with new cells. The cells in the lining of the intestines are replaced every few days.

Hundreds of thousands of new blood cells are produced every day in the bone marrow. These types of tissues are reseeded with new differentiated cells that divide from a pool of nondifferentiated stem cells. In theory, replicative senescence could reduce the number of functional stem cells. A loss of replicating stem cells could lead to deterioration in many tissues, leading to some of the symptoms of aging. In addition, senescent cells secrete enzymes that degrade tissue components and growth factors that stimulate cell division. It is possible that the presence of senescent cells within a tissue could disrupt the integrity of the tissue in many different ways.

Is there any evidence that replicative senescence leads to aging in multicellular organisms? Most of the data so far are suggestive, but not conclusive, of a cause-effect relationship. Senescent cells can accumulate in older tissues. With age, the immune system experiences a decrease in T-cell numbers and these immune cells appear to have less capacity for cell division. These declines in replicative capacity are accompanied by a reduction in telomere length. Some characteristics of aging skin can be experimentally induced by reconstituting skin *in vitro* with senescent human fibroblasts. The number of dividing cells in the lens of the eye decreases with age and these older lens cells grow poorly in culture – another characteristic of senescent cells.

As yet, there is no direct evidence that telomere length or telomere shortening affects human lifespan. However, one recent study found a correlation between telomere length and survival. People whose telomeres were shorter than average had three times the risk of developing heart disease and eight times the risk of contracting infections compared with people whose telomeres were of average length. Those with telomeres of longer than average length lived about four to five years longer than people with shorter than normal telomeres. Research in this area is only beginning and scientists do not know the significance of replicative senescence in most tissues, or even the number of senescent cells in a tissue that would be necessary to lead to aging symptoms.

It is important to note that senescence that is not triggered by telomere shortening could also be associated with aging. This kind of senescence could be triggered by various kinds of cellular damage, particularly DNA damage. DNA damage brought about by agents described in the next section, could send signals to the p53 protein, causing transient cell division arrest, apoptosis or cellular senescence.

Wear and Tear Hypothesis

It is not difficult to imagine how damage to cellular components might contribute to the aging process. External damaging agents such as radiation and chemicals, as well as internal agents such as free radicals, can seriously affect the integrity of DNA, enzymes and cellular structures. The gradual accumulation of damaged molecules and cellular components over time could affect cell function and trigger p53-mediated apoptosis and cellular senescence. A large number of age-related cellular damaging agents have been suggested to cause aging and age-related diseases such as Alzheimer's and Parkinson's. In addition, the body's ability to defend itself from these agents and to repair the damage they inflict could significantly affect the aging process. Some of the types of cellular damage that accompany aging are described below.

Many potentially damaging substances accumulate in aging cells. These originate either from external sources or from the byproducts of cellular metabolism. Chemicals such as lead, iron and aluminum, as well as some man-made chemicals such as DDT and PCBs, tend to accumulate in cells and their effects may be toxic. Damaged mitochondria and other cellular organelles, improperly processed proteins and lipids and mutated DNA accumulate in aging cells, suggesting that the ability of cells to detect, repair and replace these internal components may be defective in older cells.

One damaged cellular component that accumulates in aging cells is lipofuscin, or age pigment. This substance gradually builds up in nondividing cells such as neurons, muscle and retinal cells. Lipofuscin is a yellowish-brown pigment that collects in lysosomes, which are intracellular "waste-disposal" sites. Lipofuscin is composed primarily of crosslinked proteins and lipids. It is thought to be formed by the actions of reactive oxygen species, or free radicals, which are byproducts of normal aerobic metabolism. (Free radicals will be discussed at length below.) Although lipofuscin accumulation is considered a hallmark of aging cells, it is not known whether it is a cause of aging or simply an accompaniment to it.

Other biological materials that accumulate in cells include crosslinked, modified proteins known as "advanced glycation end products" (AGEs). AGEs are proteins to which sugars become attached at amine groups. This is followed by a cascade of reactions involving condensation, oxidation and dehydration of the modified proteins. Unlike other types of sugar modifications, called glycosylations, modifications to AGEs are irreversible. Formation of AGEs is exacerbated by the presence of free radicals; the presence of AGEs themselves increase the amount of intracellular free radicals and the crosslinking of other proteins into aggregates. Since the protein aggregates are nonfunctional and cells containing them could become less efficient in normal metabolic functions, the deposition of AGEs within cells could contribute to aging in many ways.

DNA also suffers damage in aging cells. It is estimated that the DNA within a cell's nucleus experiences tens of thousands of damaging events per day. This DNA damage is caused by external radiation, toxins, internally-generated free radicals and the spontaneous alterations that occur to DNA bases. DNA damage may occur to either the DNA phosphodiester backbone (leading to single-strand and double-strand breaks) or to the DNA bases themselves. Normal cells employ a large number of mechanisms to repair DNA damage; however, as cells age, these DNA repair mechanisms become less efficient.

The consequences of DNA damage are varied. First, DNA damage within genes can create mutations that result in the synthesis of abnormal proteins, which, in turn, can lead to cell function defects. Second, the presence of damaged DNA may trigger p53-mediated apoptosis or senescence. This could contribute to the symptoms of aging by altering the functions of, or removing, cells that make up tissues. Third, DNA damage occurring within mitochondria (the cellular organelles responsible for energy production) could lead to reductions in the number of functional mitochondria per cell, thereby reducing the metabolic efficiency of aging cells.

Experimental animals that are exposed to chronic, but sublethal, levels of high energy radiation such as gamma rays and X-rays have shorter than normal lifespans and higher levels of cancer, as well as symptoms resembling those of premature aging. Mice that bear mutations in genes whose products help to repair damaged DNA also show symptoms of premature aging. The potential link between DNA repair and aging is further suggested by several human genetic diseases characterized by symptoms of premature aging (Table 1). For example, people with Werner syndrome, Ataxia telengiectasia, Cockayne syndrome and trichothiodystrophy show degeneration of the brain and nervous system, thinning hair and other age-related symptoms. These individuals bear mutations in genes whose products detect and repair DNA damage. These premature aging syndromes, called segmental progerias, are also discussed in a later section (Do genes control aging and death?) and in the Box: Aging and Progeria Syndromes.

Table 1 Human Segmental Progeroid Syndromes

Syndrome	Genetic Defect	Mean Lifespan (years)	Symptoms
Werner	loss of function mutations in the *WRN* gene encoding a DNA helicase, defective DNA repair	47	gray hair, skin wrinkling, cataracts, diabetes, osteoporosis, heart disease, cancers
Hutchinson-Gilford	dominant negative mutation in lamin A (*LMNA*) gene, nuclear structure defects	12	skin wrinkling, loss of subcutaneous fat, hair loss, short stature, musculoskeletal defects
Ataxia telangiectasia	loss of function mutation in *ATM* gene, DNA repair defects	20	gray hair, immunodeficiency, cancers, neurodegeneration, skin atrophy, cancers
Cockayne	loss of function mutations in *CSA* and *CSB* genes, DNA repair defects	20	deafness, retinal degeneration, heart disease, UV sensitivity, cataracts, hypertension, dementia
Berardinelli-Seip	loss of function mutations in *AGPAT2* or *BSCL2* genes	40	absence of fat tissue, diabetes, hypertension, dementia
Down	chromosome 21 trisomy	60	gray hair, cataracts, neurodegeneration, loss of subcutaneous fat, vision loss, thyroid dysfunction
Trichothio-dystrophy	mutation in *XPD* gene encoding a DNA helicase, DNA repair and transcription defects	10	neurological and skeletal degeneration, wasting, brittle hair and nails

The Oxidative Stress (Free Radical) Hypothesis

One of the early hypotheses to explain aging was the "rate of living" hypothesis. This idea proposed that the lifespan of an organism is related to its energy consumption – in other words, metabolic rate determines aging rate. This idea appeared to explain the observation that animals with slow metabolic rates (such as elephants and humans) live longer than animals with rapid metabolic rates (such as mice and birds). Similarly, cold-blooded animals such as *Drosophila* have shorter lifespans at 30°C than they do at 10°C. However, the rate of living hypothesis is challenged by observations that some species do not follow the general rule. For example, bats have similar metabolic rates to mice; however, bats live ten times longer than mice. Opossums live about two years, but hummingbirds can live to be 14. In addition, the rate of living hypothesis is not supported by experimental evidence. Manipulations that alter an organism's metabolic rate do not always affect lifespan. Similarly, factors that extend lifespan, such as calorie restriction, do not necessarily affect metabolic rates.

Although the rate of living hypothesis, in its simplest form, is no longer accepted as a credible aging hypothesis, the current oxidative stress hypothesis (also called the free radical hypothesis) bears some resemblance to it. The oxidative stress hypothesis states that aging is a consequence of accumulated cellular damage caused by internally-generated reactive oxygen species, or free radicals.

Box — Aging and Progeria Syndromes

In the search for genes that influence aging, scientists are analyzing the genetic makeup of individuals who live to advanced ages, as well as individuals with genetic mutations that cause premature aging. Several human genetic diseases display symptoms resembling those of accelerated aging and some of these diseases are caused by single gene mutations. These diseases are known as segmental progeroid (or "partial early aging") syndromes. Research into Werner syndrome illustrates how studies of progeroid syndromes lead to insights about the basic mechanisms of aging.

Werner syndrome (WS) is named after German physician Otto Werner who described the syndrome in 1904. It is an extremely rare autosomal recessive disease affecting one in 25 million people. To date, only about 200 cases worldwide have been described. About three-quarters of WS cases have occurred in Japan where marriages between first or second cousins are more common than in western nations. WS can affect more than one child in a family and the gene mutation causing WS can be passed from one generation to the next.

A person with WS displays a number of symptoms resembling those found in normal aging. Children with WS appear normal until their teens, when they fail to grow significantly. As a result, WS patients are rarely over five feet tall. They soon develop gray hair, prematurely thin and wrinkled skin and calluses that ulcerate. They lose muscle and bone mass and their joints become stiff. Most WS individuals develop cataracts in both eyes by their 30s. They exhibit a range of cardiovascular problems, Type II diabetes and cancers. Despite these problems, some individuals with WS are fertile and have children. They also have normal immune systems and neurological function. The average age of death is 47, usually as a result of heart disease or cancer.

The gene that, when mutated, is responsible for WS – *WRN* – was identified in 1996. The *WRN* gene is a member of the RecQ helicase family and encodes a DNA helicase. DNA helicases are responsible for unwinding the DNA duplex prior to a number of essential functions such as DNA replication, repair and recombination. Cells that lack a functional *WRN* gene are sensitive to DNA damaging agents and accumulate an abnormally large number of mutations. In addition, WS cells in culture have a limited capacity to divide, usually undergoing fewer than 20 doublings.

What does the genetic defect in WS tell us about the mechanisms of aging? The inability of WS cells to efficiently repair DNA damage suggests that some aspects of aging may be caused by the accumulation of DNA damage. Damaged DNA sends signals through the p53 pathway, resulting in cell division arrest, cellular senescence or apoptosis. If cells that must divide, such as stem cells, cease cell division or die as a result of DNA damage, the body may have difficulties repairing tissues. Cells isolated from WS patients do not divide as many times in culture as cells from normal people, suggesting that they experience premature replicative senescence, perhaps due to premature erosion of their telomeres. This may indicate that aging is caused by replicative senescence, at least in some cell types. In 2004, a knockout mouse model for WS was developed. The WS knockout mouse lacks both *WRN* alleles as well as the genes encoding the enzyme telomerase. WS mice develop premature aging symptoms similar to those seen in humans. The hope is that these WS mice will allow even further exploration of the genetic causes of aging.

Figure 4

Reactive oxygen species (free radicals) are generated during aerobic metabolism. Superoxide dismutase (SOD), catalase, peroxidases and other antioxidants help neutralize free radicals and protect cells from their damaging effects.

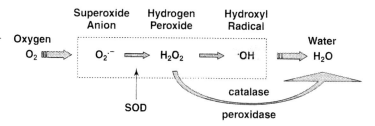

Free radicals are unstable, highly reactive atoms or molecules with one or more unpaired electrons and are created as normal byproducts of aerobic energy metabolism. Free radicals include superoxide anions ($O_2^{.-}$), hydroxyl radicals ($^.OH$) and hydrogen peroxide (H_2O_2) (Figure 4). Free radicals can damage almost all cellular components, from membranes to DNA and proteins. They attack nucleic acid bases, protein side-chains and the double bonds of unsaturated fatty acids. Free radicals contribute to the formation of crosslinked proteins and lipids such as the AGEs and lipofuscins, as described above. Cellular membranes that are damaged by peroxide become rigid and lose permeability. If the cell cannot repair the damage, cellular functions may be impaired or the cell could be forced down biochemical pathways leading to senescence or apoptosis.

Free radical damage is counteracted by the cell's natural antioxidant defenses, as well as its ability to repair DNA, proteins and lipids. The antioxidant enzymes superoxide dismutase, glutathione peroxidase and catalase act by scavenging free radicals and hence protecting cells from the damaging effects of free radicals. Some studies show correlations between cellular levels of enzymatic antioxidants and lifespan. Transgenic flies with increased superoxide dismutase and catalase activities show a 15 to 30 percent increase in lifespan and reduced levels of free radical damage. One study showed that transgenic expression of human superoxide dismutase in *Drosophila* neurons may contribute to a 40 percent increase in lifespan. Small antioxidant molecules also protect cells from free radicals. These include ascorbate, urate, tocopherols, flavonoids and carotenoids – many derived from dietary sources, particularly fresh fruits and vegetables. The role of these substances in oxidative damage and aging is unclear. Results of studies on dietary supplementation with antioxidants in laboratory animals have been mixed – sometimes extending lifespan and other times having no effects.

So far, the strongest support for the oxidative stress hypothesis is a general correlation between the rate of free radical production and the lifespans of animals. For example, rats have a five-fold shorter lifespan than pigeons and a correspondingly higher rate of free radical production. The long-lived rodent *Peromyscus leucopus* (white-footed mouse) contains lower levels of free radicals, higher levels of cellular antioxidant enzymes and less protein oxidation damage than its shorter-lived cousins *Mus musculus* (common house mouse). In addition, the amount of free radical-mediated damage appears to rise in aging cells. Some experimental evidence for the oxidative stress hypothesis exists. In mice whose lifespans are increased by calorie restriction, the rates of free radical generation are lower, and the amount of free radical damage is less than in their normal counterparts. Mice with an engineered inactivation of the gene encoding *p66*^{Shc} (which normally induces apoptosis in response to oxidative stress) have one-third longer lifespans and enhanced resistance to free radicals.

If free radical damage is severe, cells may undergo apoptosis, perhaps as a side-effect of p53-sensed DNA damage. In addition, there may be a relationship between free radical damage and telomere-triggered senescence. Human fibroblasts grown in low oxygen have a longer lifespan than those grown in higher oxygen concentrations. The cells grown in higher oxygen show rapid telomere shortening prior to cellular senescence.

Although there is a general correlation between free radical-mediated damage and aging, and some recent studies suggest a direct role for free radical damage in aging, further research is required before a cause-effect relationship can be firmly established.

DO GENES CONTROL AGING AND DEATH?

Most scientists now agree that aging is not genetically programmed, in the sense that one or more genes provide a mechanism and a timetable for the aging process. The reason for this is that there is not a way in which evolution could select for genes that directly cause aging. Natural selection is thought to operate only on genes that affect reproduction. Once organisms reproduce successfully, evolution is indifferent to the actions of genes. Because of this, genes whose effects are important for bringing the organism to the age of reproduction, but whose effects may be detrimental after reproductive age, can become fixed in a population. However, it is also clear that genes influence all aspects of aging. It is estimated that about 25 percent of the variation in lifespans between individuals is linked to genetics. The remainder of the variation is due to differences in behavioral and environmental factors such as diet or smoking.

Evidence that genes influence the progress of aging comes from several observations. The lifespans of identical twins are closer to each other than those of either fraternal twins or nontwin siblings. Identical twins die about 36 months apart, fraternal twins die about 75 months apart and siblings die about 106 months apart. The offspring of people who live into their 100s (centenarians) are also likely to have long lifespans. Each species has a characteristic, and heritable, maximum lifespan. Mice live about 2 years under protected conditions; humans live to a maximum of about 120 years. Large-bodied breeds of dogs have maximum lifespans of under 10 years; whereas smaller breeds can live beyond 15 years. The fact that lifespans of different species that occupy similar ecological niches are different, and these differences are stably inherited within the species, supports the idea that aging is genetically influenced.

The concept that aging is *influenced* by genes, but that aging is not *caused* by genes is a subtle but important one. If aging were programmed by genes, we would expect to find specific aging genes, conserved throughout evolution, which, when inactivated, would abolish aging. No such aging genes have been found. The genes that influence aging affect various biological traits that contribute to the rate and symptoms of the aging process. It is estimated that up to 7,000 genes may influence the pace or manifestations of aging, but no one gene is responsible for all aspects of aging. Aging is a multifaceted phenomenon that appears to act differently in each species and varies in rate and symptoms between individuals within a species.

This multigenic influence upon aging is well illustrated by human diseases known as "segmental progeroid syndromes." Individuals with these genetic diseases show symptoms resembling various aspects – but not all aspects – of aging (Table 1). Many of these diseases are caused by mutations in single genes. (See Box: Aging and Progeria Syndromes.)

The complicated and multifaceted effects of genes on aging is also seen in experimental animals. About 40 genes in the nematode worm *C. elegans* affect its lifespan, and many of these genes influence hormone-regulated metabolism or enhance resistance to free radical-mediated damage. For example, mutations in the *age-1*, *daf-2* and *daf-16* genes significantly affect the worm's lifespan. The product of the *daf-2* gene (DAF-2 protein) is an insulin-like receptor protein that is found within cells. Its normal function is to interact with extra-cellular insulin-like hormones that are released as a result of food availability or stress; it then sends signals to a number of other proteins within the cell (including the *age-1* gene product). These, in turn, send signals to the DAF-16 protein. DAF-16 regulates the synthesis of more than 300 gene products involved with oxidative stress defenses, protein degradation and antimicrobial responses. Worms with mutations in the *daf-2* and *age-1* genes have long lifespans (twice that of normal worms) and are resistant to many stresses induced by free radicals, heat, UV light and heavy metals. However, mutations that inactivate the *daf-16* gene result in reduced lifespans. Another gene, *clk-1*, appears to be involved in free radical generation in the cytoplasm. When the *daf-2* and *clk 1* genes are simultaneously inactivated, the double-mutant worms live up to five times longer than their wild-type counterparts (Figure 5).

The complexity of genetic effects upon aging is illustrated further by the actions of the *sir2* gene. Yeast cells that bear extra copies of the *sir2* gene have longer than normal lifespans. The Sir2 protein is a deacetylase – an enzyme that removes acetyl side groups from proteins and, by doing so, regulates the ways in which these proteins control the activities of other genes and proteins. For example, when the Sir2 protein binds to yeast genes that code for ribosomal RNA (an essential component of ribosomes and, hence, cellular protein synthesis), it prevents these ribosomal RNA genes from being transcribed. This, in turn, reduces the toxic effects that result from age-related excesses of ribosomal RNA. In *C. elegans*, the gene that corresponds to the yeast *sir2* gene, *sir2.1*, encodes a protein that affects metabolism by repressing the cellular responses to an insulin-like hormone. In mammals, the gene that corresponds to yeast *sir2*, *SIRT1*, encodes a deacetylase that represses the activity of the p53 protein. This repression of p53 activity may reduce the amount of apoptosis that occurs in response to cellular damage, thereby preserving the integrity of tissues over time. In yeast, Sir2 activity is regulated by the relative concentrations of nicotinamide adenine dinucleotide and nicotinamide, which are produced during metabolism. Hence, Sir2 links metabolic state, gene expression, hormone signaling and apoptosis with aging.

The complex relationships between aging and hormones is illustrated by genes that affect aging in flies and mice. *Drosophila* bearing mutations in genes whose products control physiological responses to insulin and the insulin-like growth factor IGF-1 are dwarfs, but have lifespans that are double those of wild-type flies. In contrast, mice with gene mutations making them defective

Figure 5
Adult lifespans of wild-type and mutant lines of *C. elegans*. Organisms with mutations in both *daf-2* and *clk-1* genes have significantly longer lifespans than organisms with mutations in only one of these genes. Source: Hekimi, S. and Guarente, L. 2003. Genetics and the specificity of the aging process. *Science* **299**:1351.

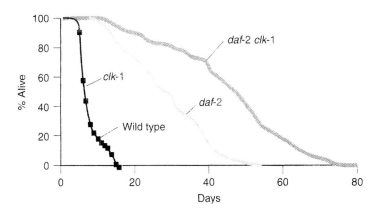

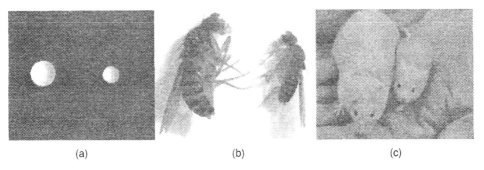

(a)   (b)   (c)

Figure 6

Yeast, *Drosophila* and mice that bear mutations in genes affecting glucose or insulin/IGF-1-like signaling are dwarfs and have extended lifespans. (a) Yeast with mutations in the *sch9* gene (right) grow more slowly and live three times longer than wild-type yeast (left). (b) Flies with mutations in the *chico* gene (right) live up to 50 percent longer than wild-type (left). (c) GHR/BP mice (right) are deficient in IGF-1 and live 50 percent longer than wild-type mice (left). Source: Longo, V.D. and Finch, C.E. 2003. Evolutionary medicine: from dwarf model systems to healthy centenarians? *Science* **299**:1342-45. (Fly image provided by D. Gems. Mouse images provided by A. Bartke.)

in their responses to IGF-1 are also dwarfs, but have extended lifespans. Mice with homozygous mutations in the *Prop-1* and *Pit-1* genes have 65 percent longer lifespans than their wild-type counterparts. These mice also have lower than normal levels of IGF-1, growth hormone, thyroid stimulating hormone and prolactin – all of which affect normal growth and metabolism (Figure 6).

Recent research also suggests links between aging and genes that control cellular senescence and apoptosis. When mice are engineered to have inactive p53 genes, they show a 20 percent decrease in lifespan and an increase in cancer susceptibility. Interestingly, mice that synthesize too much p53 also have a decreased lifespan, even though they are resistant to cancer. These mice show symptoms of premature aging, including weight loss, osteoporosis, delayed wound healing, organ atrophy and impaired regeneration of tissues following stress. It is possible that, due to loss of cells following apoptosis, p53-mediated increases in apoptosis or cellular senescence could explain the organ atrophy seen in these mice.

Do these same genes operate in humans to extend or shorten lifespan? There is some indirect evidence that this may be the case. For example, humans with pituitary gland abnormalities may secrete excess growth hormone. This results in enlarged body tissues such as bones and reduced lifespans due to heart disease and cancer. However, whether this lifespan reduction is due to aging or to increases in specific diseases is not clear. Interestingly, growth hormone deficiencies in humans also lead to reduced lifespan and are associated with health problems such as early atherosclerosis and loss of muscle and bone mass. Since it is not possible to do gene manipulation experiments in humans, it is still not clear whether the genes and metabolic pathways that affect the lifespans of model organisms also have the same effects in humans.

Questions About the Search for Immortality

DO ALL ORGANISMS AGE? ARE ANY ORGANISMS IMMORTAL?

No organism is immortal. Disease, predation, starvation or accident will eventually claim every living thing, no matter how long-lived. However, aging is not universal.

Bacteria, which reproduce by simple binary fission, do not age. There is no discernable parent or offspring for these prokaryotes and they do not appear to suffer a progressive loss of function or fertility over time. However, just being unicellular is not necessarily a way to escape aging. *Saccharomyces cerevisiae* (brewer's yeast), which reproduces by budding, experiences a form of aging. Mother yeast cells can only bud for a limited number of times before undergoing senescence. Conversely, being multicellular is not necessarily a ticket to inevitable aging. *Hydra*, which reproduce by an asexual budding process, do not appear to age.

Interestingly, there are some higher eukaryotic organisms that appear to age very slowly, if at all. These include certain fishes, tortoises, amphibians and the American lobster. A recent survey revealed that about 16 percent of Yelloweye rockfish harvested in the waters off Alaska were over 100 years old. The Calico rockfish can live to 205 years with no apparent signs of aging. The red sea urchin, found in the waters of British Columbia, can exceed 200 years old, and remain fertile. The giant Sequoia can live for over 2,000 years and the bristlecone pine can live for over 4,000 years (Figure 7). Although very little is known about the biology of these long-lived species, a common feature is that they grow steadily and do not reach a fixed adult size. Research on the biology of these long-lived organisms may reveal clues about the mechanisms of aging.

COULD HUMANS BECOME IMMORTAL? IS THERE AN UPPER LIMIT TO THE HUMAN LIFESPAN?

Like all living things, humans are mortal. Even if we did not suffer from aging, we would eventually die from accidents, infections, homicides and suicides. Interestingly, in the industrial world, motor vehicle accidents kill more people than Alzheimer's disease. It has been calculated that a theoretical human population that died only as a result of accidents would have a life expectancy of about 1,200 years.

At present, the oldest documented age for a human is 122 years (See Box: Tale of a Supercentenarian). The claims that people in the Caucasus Mountains of the Republic of Georgia live to the age of 150 have been disproved. Similar claims for longevity of people in the village of Vilcabamba in Ecuador have also been discredited.

Although human *average* lifespans, at least in industrialized countries, have increased significantly over the last 100 years, it is still not clear whether the maximum lifespan may also be increasing correspondingly. Some data appear to suggest that maximum lifespans may be increasing. Although centenarians are still rare, their numbers are increasing. In 1950, there were 2,300 people over 100 in the United States. This number has now increased to about 40,000. At

Figure 7
The giant Galapagos tortoise (*Geochelone elephantopus*) lives to almost 200 years. This endangered tortoise grows to about 5 feet in length and can weigh as much as 550 lbs. The bristlecone pine (*Pinus aristata*) is one of the longest-living organisms on Earth: it can live to over 4,000 years. It is a slow-growing tree, found at high altitudes. Source: Science Photo Library

present, the numbers of centenarians are doubling every 10 years. Demographers predict that up to 4 million centenarians will be alive in the United States in 2040.

But do these statistics indicate an increase in maximum lifespan? The answer to this question is highly controversial. Some researchers feel that the increase will continue, giving humans a maximum lifespan of 150 years by the middle of the next century and an average lifespan of 120. Other researchers are less optimistic. They believe that science has conquered most of the diseases and environmental threats to younger people and it is now left with the more daunting task of conquering diseases of the aged. Even in the absence of age-related diseases, cells and tissues would still age, resulting in organ failures that could not be ascribed to specific diseases. The pessimists explain that the increase in the number of centenarians during the last century is the result of the same factors that raise the average life expectancy. There are now more humans surviving to the ranks of the elderly and, hence, a larger pool from which centenarians can emerge. In addition, the centenarians of today were born and grew up at the beginning of the era of sanitation, antibiotics and medical advances. This beneficial environmental milieu has allowed more people born at the beginning of the last century to survive to old age. Also, the protected environments of hospitals and institutions allow more older people to attain the maximum human lifespan. However, some scientists argue, there is still a maximum lifespan – about 120 years – and this will remain unchanged until science can arrest the fundamental biological causes of aging.

It may seem paradoxical, but the same scientists who declare that there is a maximum human lifespan of about 120 years also assert that we will someday dramatically extend this maximum. They predict that science will make major advances in understanding the mechanisms of aging and that these will translate into life-extending technologies within the next 20 to 50 years. If we could slow the aging process, they argue, we could see human lifespans of hundreds of years, with corresponding increases in health and vitality into advanced years. Scientists can now extend the lifespans of laboratory organisms such as yeast, worms and mice from two to six times longer than normal (See Box: The Quest for Methuselah Mouse). These advances have resulted from diverse genetic and behavioral modifications, as discussed below. Whether any of these manipulations will work in humans remains to be determined.

Even though somatic cells are mortal, germ cells have the capacity to live forever. The germ cells that manufacture eggs and sperm are passed from generation to generation with none of the accumulated deterioration that is characteristic of aging. If research can discover the features of germ cells that make these cells immortal, it may provide hints about why somatic cells are mortal. Another type of human cell – the cancer cell – is also theoretically immortal. Cancer cells do not undergo replicative senescence, but divide indefinitely, at least in laboratory situations.

Box — The Quest for Methuselah Mouse

The quest began in the fall of 2003, when Dr. Aubrey de Grey of the University of Cambridge announced the inauguration of the Methuselah Mouse Prize – a contest designed to reward scientists for producing the world's oldest mouse. Set up in the manner of the recent X Prize, which triggered the development of private space travel for the public, the Methuselah Mouse Prize aims to create public interest and support for antiaging research. The ultimate goal is to stimulate scientific research that will translate into effective longevity treatments for humans. At present, the Methuselah Mouse Prize stands at over $500,000 and six research teams have entered the competition.

The contest rules are simple. Scientists can use any type of intervention – genetic, nutritional or behavioral – that results in record-breaking lifespans for the laboratory mouse, *Mus musculus*. There are two subtypes of the prize: the Postponement Prize is given for the mouse that lives the longest and the Reversal Prize is given for the oldest mouse whose longevity interventions began later in life.

The first Methuselah Mouse Postponement Prize was awarded to Dr. Andrzej Bartke of Southern Illinois Medical School for his mouse GHR-KO 11C. 11C lived six days less than five years, equivalent to about 180 human years. A normal laboratory mouse lives about two years. 11C was a member of the Laron mouse strain that was genetically engineered to lack the gene encoding the growth hormone receptor. As a result, Laron mice cannot respond to growth hormone and are dwarfs. They have other defects, such as decreased levels of insulin-like growth factor I (IGF-1), delayed sexual maturation and high postnatal mortality. The next winner of the Methuselah Mouse Postponement Prize will need to produce a mouse that lives longer than GHR-KO 11C.

At present a winner for the Reversal Prize has not been announced, although the current record holder is Dr. Tom Kirkwood for this mouse "Charlie" that lived four years on a calorie restricted diet.

Other mice have survived to about four years of age. Yoda, who died on April 22, 2004 lived to the age of four years and 12 days – equivalent to about 150 human years. Like GHR-KO 11C, Yoda was a dwarf mutant mouse with mutations that affected insulin and growth hormone production and the functions of his pituitary and thyroid glands. As a result, he was about one-third the size of a normal mouse, frail and sensitive to cold. Yoda was housed in a special pathogen-free facility at the University of Michigan Medical School and was part of a research project directed by Dr. Richard A. Miller. Yoda remained mobile and sexually active into old age and was housed with a larger female mouse, Princess Leia, whose body warmth kept him from freezing to death. He did not suffer from diseases of old age and the cause of his death is uncertain.

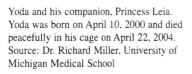

Yoda and his companion, Princess Leia. Yoda was born on April 10, 2000 and died peacefully in his cage on April 22, 2004.
Source: Dr. Richard Miller, University of Michigan Medical School

Box (continued)

Although Yoda and GHR-KO 11C were dwarf mice with defects in insulin signaling and growth hormone action, normal mice kept on severe calorie restricted diets can also live to advanced mouse ages. Scientists are eager to learn the biochemical basis for how calorie restriction and insulin metabolism contribute to longevity in mice and other organisms such as *C. elegans* and *Drosophila*. The impetus of the Methuselah Mouse Prize may speed the race towards this understanding and towards effective antiaging interventions for humans.

Paradoxically, the study of cancer, one of the major killers of older mammals, may shed light on the fundamental mechanisms of aging.

IF NOT IMMORTAL, CAN WE EXTEND OUR LIFESPANS SIGNIFICANTLY?

As mentioned previously, many scientists claim that we will see major increases in human longevity within the next 50 years. These predictions are based upon recent advances in understanding the mechanisms of aging, combined with laboratory manipulations that have extended the lifespans of certain model organisms such as *Drosophila*, mice and worms.

Dozens of genes, when mutated, extend or reduce the lifespans of laboratory animals. These were discussed in the section *Do genes control aging*? It is thought that these genes affect the aging process by affecting fundamental cellular processes such as the rate of free radical generation during metabolism and the efficiency of intracellular DNA repair – processes that modulate aspects of the aging process. Although it is possible that the future will see genetic manipulation of humans for the purposes of specific age-related disease treatments, it will be a very long time before this could be entertained as an antiaging therapy. Gene therapy that would eliminate aging would require genetic modification of every cell in the body, and the only efficient way to introduce genetic changes into all parts of an organism is to alter the germ cells that create the next generation. At the present time, genetic modification of the human germ line is considered both technically difficult and ethically unjustifiable.

The only non-genetic manipulation that is known to extend the average and maximum lifespans of organisms is calorie restriction. Calorie restriction is a diet that contains 30 to 50 percent fewer calories than an ad lib diet, while retaining a balance of nutrients. Animals kept on calorie restricted diets often have longer, healthier lives. The effects of calorie restriction in humans are now being assessed in clinical trials, and possible "calorie restriction mimetics" – drugs that mimic the effects of calorie restriction – are now being tested (See Box: Calorie Restriction and Longevity).

Is it possible that delayed reproduction could lead to an increase in the human lifespan? As described previously in this booklet, *Drosophila* that are selected for late reproduction over many generations develop longer lifespans. One study suggests that a similar process may occur in humans. The authors of this study discovered that women who lived beyond the age of 100 were four times more likely to have become mothers in their 40s than those who died at the age

Box — Calorie Restriction and Longevity

The only nongenetic intervention that is documented to extend both average and maximum lifespans of many species is calorie restriction (CR). By reducing calories from 30 to 50 percent below ad lib eating levels, animals such as mice, dogs, rats, flies, fish and worms can gain significant extensions in lifespan. For example, mice on CR diets live an average of 40 percent longer than mice who eat freely. This would translate into an average human lifespan of about 112 years. Laboratory animals on CR diets have lower incidences of age-related diseases such as cancer, diabetes, cardiovascular disease, kidney disease, osteoporosis and neurological conditions. Older CR animals have youthful levels of blood sugar, lipids, insulin, and blood pressure, and perform like younger animals in learning tests.

The first major clinical trial of CR in humans was initiated by the National Institute on Aging in 2003. Researchers will monitor the health of people on a 10 percent calorie reduced diet. Results will be available in 2007. Two other clinical trials will measure blood pressure, cholesterol levels and various disease rates in CR patients. Two trials of CR in monkeys are now half completed (monkeys have a lifespan of about 50 years). So far, the data are encouraging, since these animals show the same physiological effects as those observed in mice and rats.

Despite benefits, CR should be undertaken with caution. Rapid adoption of CR can damage health; therefore, it must be phased in over a long period – equivalent to about 2.5 human years. CR must be monitored so that nutrition is not compromised. Unless there are adequate levels of all nutrients, CR will simply become malnutrition. Children should not undertake CR, since it stunts growth and sexual development. Other negative effects reported by people on CR include anemia, loss of bone mass, chills, depression, slow wound healing and reduced energy levels. People on CR can lose muscle strength, are often hungry and have a reduced interest in sex. A common joke about CR is that it may not extend your life, but will certainly make it seem longer.

How does CR work? So far, no one knows. At present, the most accepted hypothesis is that CR places the organism into a "stress response" state with enhanced cellular repair mechanisms. In the wild, this state would keep an animal healthy until food supplies returned to normal and they could reproduce.

Because most humans would be reluctant to follow a CR diet, scientists are attempting to find a drug, or "CR mimetic," that will have the positive effects of CR without the negative ones. Some interesting candidates have emerged recently and are now being tested. One is resveratrol, a chemical that is abundant in red wines. In yeast, flies, and worms, resveratrol extends lifespan up to 70 percent. This finding is interesting, since modest

Source: Getty Images

consumption of red wine has been associated with beneficial effects on cancer and cardiovascular disease rates. The dosage effects of resveratrol appear to be critical; at high dosages, the chemical may reduce longevity. Other possible CR mimetics being tested are 2-deoxyglucose, which inhibits sugar metabolism and hence, energy and free radical levels, and metformin, which affects insulin activity. Scientists predict that a useful CR mimetic may be available within ten years.

of 73. They concluded that pregnancy after the age of 40, as well as later menopause, correlates with longevity. Some scientists speculate that the trend towards later reproduction in modern industrial societies may, after many generations, select for a longer-lived population similar to the effects seen in fruit flies. If so, a genetic selection for longer-lived humans, based on later reproduction, would take several hundred years to become apparent.

DO DIETS, SUPPLEMENTS AND LIFESTYLES AFFECT HUMAN LIFESPAN?

Probably since the beginning of human consciousness people have sought remedies to extend their lives. As a result, purveyors of antiaging elixirs and lifestyle modifications have developed a multi-billion dollar industry supplying products and advice to those seeking youth and longevity. Unfortunately, none of the thousands of claims made for antiaging products have shown any verifiable effects on human aging or lifespan. Despite promising research data from studies of laboratory animals, there is simply no intervention – chemical, behavioral, dietary, medical or genetic – that has been shown to slow aging or increase the maximum lifespan of human beings. Medical science can offer many interventions that reduce the likelihood of acquiring certain diseases. For example, exercise and low fat diets can reduce the incidence of cardiovascular disease and some cancers. However, these do not affect the inexorable process of aging. In the absence of one disease, the aging body will simply succumb to the effects of another age-related condition.

In 2002, a group of 51 scientists involved in aging research published a statement called "The Truth about Human Aging" in order to warn the public about pseudoscientific antiaging products (see **www.sciam.com/agingstatement.cfm**). In this position paper, they state,

"Since recorded history individuals have been, and are continuing to be, victimized by promises of extended youth or increased longevity by using unproven methods that allegedly slow, stop or reverse aging. Our language on this matter must be unambiguous: there are no lifestyle changes, surgical procedures, vitamins, antioxidants, hormones or techniques of genetic engineering available today that have been demonstrated to influence the processes of aging. We strongly urge the general public to avoid buying or using products or other interventions from anyone claiming that they will slow, stop or reverse aging."

Despite this pessimistic news, there are some human clinical trials in progress to assess the possible benefits of certain antioxidant and vitamin supplements in relieving a number of age-related conditions such as cardiovascular disease and macular degeneration.

Several studies of antiaging therapies show promising results in laboratory animals. Certain synthetic versions of superoxide dismutase and catalase enzymes alleviate oxidative stress disorders in mice and extend the longevity of *C. elegans*. The drug 4-phenylbutyrate (PBA) extends the maximum lifespan of *Drosophila* by 50 percent. PBA may work by switching on dozens of genes, including one that encodes superoxide dismutase. A substance found in plants and abundant in red wine, resveratrol, appears to extend the lifespans of yeast, *Drosophila* and *C. elegans* (See Box: Calorie Restriction and Longevity). Another substance, acetyl-L-carnitine, in combination with the antioxidant lipoic acid, appears to improve the energy levels and mental capacities of aging rats by increasing mitochondrial respiration and scavenging the resulting free radicals. It is still unclear whether this combination makes the rats live longer. The antioxidant N-tert-butyl-α-phenylnitrone (PBN) increases the lifespan of a strain of rapidly-aging mice from 42 to 56 weeks. This substance also improves the short-term memory of aging gerbils. The National Institute on Aging is now testing aspirin (an antioxidant and anti-inflammatory), 4-hydroxy-PBN (an antioxidant) and nitroflurbiprofen (an anti-inflammatory) for longevity effects in mice. The results should be available within several years.

Although there is no evidence yet that any substance or treatment can slow the aging process in humans, many scientists are optimistic that effective antiaging treatments will be designed based on an understanding of the biological mechanisms of aging. Some predict that one or more of these treatments may be available within 10 to 20 years. In the meantime, healthy eating, regular exercise and avoiding tobacco and sun exposure may be the most effective ways to deal with aging and age-related diseases.

WHAT IS THE FUTURE FOR ANTIAGING RESEARCH?

The future looks bright for antiaging research. And the predictions of some researchers sound positively utopian.

As described previously, genes affect longevity (either directly or indirectly) by influencing some of the multitude of biological factors that cause aging and by affecting the predisposition to certain age-related diseases. The challenge will be to identify those genes in humans that most directly influence aging and to design interventions based on the actions of these genes. One of the most promising approaches to finding human aging genes is to study the genetic makeup of people who suffer from premature aging syndromes and those who live to extremely old ages (See Box: Aging and Progeria Syndromes). Several centenarian studies are now in progress. The European Commission is sponsoring "The Genetics of Healthy Aging" project. This study will analyze the health, lifestyle and genetic makeup of 2,800 pairs of siblings over the age of 90 and compare the results with data from a younger control population. Results may be available by 2010. Several smaller studies have already identified certain gene alleles that are associated with longevity. Many centenarians express genes that help control the inflammatory response, perhaps leading to greater resistance to the damaging effects of chronic inflammation. Many carry a certain allele of the gene encoding apolipoprotein E – *ApoE2* – the product of which is part of complexes that contain cholesterol. This allele may help protect these individuals from cardiovascular disease or Alzheimer's. Information gleaned from the Human Genome Project is being used to search the entire genome for human longevity genes. One such study has identified variants of a gene known as *MTP* and has linked these variants with longevity. The products of this gene may be involved in lipoprotein synthesis. It is still unclear whether these genes directly influence the aging process, or if they simply affect age-related diseases such as cardiovascular and Alzheimer's disease.

In the future, could we use gene therapy to arrest aging? Scientists now accept that aging is a multifaceted process resulting from accumulated damage to the molecules of life – DNA, proteins, lipids and carbohydrates – as well as from the effects of cellular processes such as replicative senescence and apoptosis. Because aging is so complicated, it is likely that a vast number of genes affect aging. Even if scientists could successfully introduce genes into all cells of an adult human or knock out the function of other genes, it seems unlikely that a sufficient number of genes could be engineered to dramatically affect the aging process. It would also be extremely difficult to do this without upsetting the fine balance that exists between multitudes of gene products, some of which affect many basic physiological processes such as growth, development and metabolism. We already know that gene manipulation in model organisms such as *Drosophila* and *C. elegans* can affect longevity, but that it can also cause undesirable side effects. Despite present technical and ethical limitations of the use of genetic engineering, some scientists believe that a degree of somatic gene therapy may become a possibility in the distant future, if not to combat aging itself, then to treat specific diseases associated with aging.

Even if gene therapy does not become a serious approach to human antiaging, it may be possible to screen people for their genetic makeups and then custom design interventions that address each individual's specific genetic predispositions. For example, it may be possible to detect certain alleles of genes that affect lipid metabolism or DNA repair – and then predict a person's susceptibility to cardiovascular disease or cancers. Drug therapies, medical interventions or behavioral modifications may then extend that person's lifespan by addressing their particular disease susceptibilities.

Research scientists and biotechnology companies are also pursuing calorie restriction mimetics, as described previously (See Box: Calorie Restriction and Longevity). If any of these prove effective in both laboratory animals and human clinical trials, they may be available for use within 10 to 20 years.

Antiaging researchers predict that regenerative medicine will combat some of the symptoms of aging. In this scenario, cloning and stem cell technologies will provide abundant transplantable tissues to allow aging organs and tissues to be replaced, thereby curing age-related diseases and extending healthy lifespan (Figure 8). Scientists agree that stem cell technologies could one day allow physicians to replace some worn-out tissues or organs. Stem cells are undifferentiated cells that retain the potential to divide indefinitely and to differentiate into many other types of cells – including bone, muscle, nerve, cartilage, and immune system cells. Stem cells are found in very early embryos and in some adult tissues such as bone marrow. It is possible to isolate embryonic or adult stem cells and grow them for long periods in culture dishes. Scientists are now attempting to stimulate these cultured stem cells to differentiate into the hundreds of specialized cell types that could be used either to replenish damaged tissue or to grow into replacement tissues in culture. Researchers have now been able to transform embryonic stem cells into heart muscle cells that will grow into heart tissue when implanted into mice. Mice with spinal cord injuries regain some mobility after they are injected with human stem cells. As yet, scientists cannot grow replacement tissues from human stem cells in culture dishes. However, in the future, this may be possible.

Cloning transgenic animals for the purposes of organ transplants is at least theoretically possible, as is cloning a patient's cells, in order to grow them into replacement tissues that are immunological matches. This latter form of therapy is called "therapeutic cloning" and is highly

Figure 8
In the future, replacement organs and tissues for defective aging organs may be derived from cloned animals or from stem cell technologies. Source: Getty Images

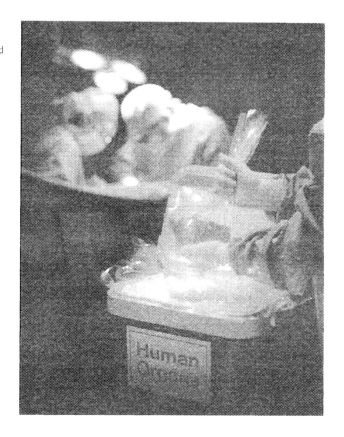

controversial. In this method, a nucleus from a patient's cell would be transplanted into an enucleated human egg and the egg would be stimulated to form an early embryo – the same procedure used to create Dolly the sheep and other cloned animals. At the blastocyst stage (about 200 cells), the embryo would be disrupted and the stem cells would be harvested. These cells would then be grown and differentiated in a culture dish to create replacement tissues. Because an embryo is destroyed in the process, therapeutic cloning introduces the ethical problems regarding when life begins.

The possibility of using regenerative medical technologies to repair or replace body parts has exciting implications for the correction of many defects, including those exacerbated by the aging process. Unfortunately, it is unlikely that regenerative medicine could replace much of the human brain without altering personality, memory and individuality. Neurons function by maintaining dozens of specific connections with other neurons. In order to repopulate a damaged brain with new neurons, the new neurons would have to make the same connections to the same neurons that the original neurons made. Unless these limitations can be overcome, the human brain may remain the weak link, subjecting us to age-related neurological diseases and dementias, even if our bodies remained young.

Another futuristic prospect to arrest or cure aging is nanotechnology. Nanotechnology is based on the use of machines the size of individual atoms and molecules. (A nanometer is one-billionth of a meter and the size an atom.) Some foresee a day when computer-controlled

Box — Cryonics and the Dream of Immortality

Ted Williams, the legendary Boston Red Sox star, suffered from heart failure and was pronounced dead on July 5, 2002. His journey, however, had just begun. Within minutes, Williams was rushed to a local funeral home where he was packed in ice, injected with a number of drugs, then quickly shipped to Scottsdale, Arizona via chartered jet. Once his body arrived at the Alcor Life Extension Foundation, he was brought to an operating room and connected to a heart-lung machine; his blood was replaced with cryoprotectants and his body temperature was gradually reduced to -196 degrees centigrade. He now resides upside down in a large stainless steel thermos bottle filled with liquid nitrogen. Williams is one of about 100 people who have undergone cryonic freezing and are stored in liquid nitrogen, waiting for future rejuvenation.

Cryonics is defined as the practice of freezing the body of a person who has just died, with the hope of future resuscitation in an era when technology will cure the cause of death and return the person to life. The cryonics movement began in 1964 after the publication of the book "The Prospect of Immortality" by Robert Ettinger, a physics teacher and science fiction writer from Michigan. Commercial cryonics services appeared in the 1970s. At present, there are five organizations that offer cryonic preservation – all of them in the United States. The largest of these, Alcor, charges $120,000 for a whole-body preservation and $50,000 for a head-only ("neuro") preservation. Alcor members may also have their pets frozen and store some of their important possessions for use after rejuvenation.

Could cryonics work? Most scientists say no. At present, there is no evidence that whole animals can survive a freeze-thaw cycle. Cells and tissues are damaged by ice crystals that create holes between cells and disrupt cellular structures and membranes. The smallest life forms that can survive freezing are eggs, sperm, early stage embryos and tiny insects. New techniques, including vitrification that freezes tissue without forming ice crystals, could be more effective; however, little scientific research has been done using this method.

Cryonics supporters admit that the possibility of future resuscitation is merely speculative. They place their faith in a future that they foresee as a time of great wealth, remarkably advanced technology and utopian peace. Beyond

Bigfoot dewar flasks at Alcor Life Extension Foundation. Each vacuum-insulated container, filled with liquid nitrogen, holds up to four whole patients, or ten neuro-patients. Source: Alcor Life Extension Foundation nanomachines, or nanobots, will travel through the body, seek out damaged cells and molecules, clean out arteries, kill cancer cells and robotically replace defective mitochondria. In one futuristic prediction, age-damaged circulatory systems could be lined with unbreakable synthetic materials, extending the human lifespan by eliminating heart disease, strokes and by resisting the invasion of cancer metastases. Optimistic predictions state that some types of nanotechnologies may be available within 30 years.

People's lives are already being extended by the use of artificial organs. For example, kidney dialysis machines have been in use since 1948. At present, left ventricular assist devices are in clinical use as a bridging therapy prior to heart transplantation. Totally artificial hearts are currently undergoing clinical evaluation. Beyond artificial organs are the new bioartificial devices (or "bionic" organs). Bioartificial organs are mixtures of living cells or tissues embedded in mechanical scaffolds that allow the tissues to perform the biochemical functions of the normal organ. Several bioartificial kidneys have been developed and consist of dialysis filters mixed with live renal proximal tubule cells. These devices are now in clinical trials. Bioartificial livers are also in human clinical trials. It is conceivable that many human organs could be replaced by bionic components within the next few decades.

A question that remains, however, is whether any of these regenerative medical breakthroughs would significantly extend the lives of most people. Unless all body components, including the human brain, could be replaced, aging may still claim its victims by attacking those body structures that could not be replaced. Since replacement or regeneration of the human brain would probably alter those qualities that we consider "self," aging of the brain may be the limiting factor in creating an immortal bionic individual.

Some people who are unwilling to wait 30 to 100 years for genetic engineering and regenerative medicine to cure their aging bodies are placing their hopes on another procedure – cryonics. Cryonics encompasses a number of techniques that freeze the body of a person who has just died, in order to preserve it for a future time when there may be a cure for the disease that killed them. The idea is that cryogenically preserved people will be thawed, revived, cured of their diseases and returned to a rewarding life (See Box: Cryonics and the Dream of Immortality).

WHAT'S THE BEST WAY TO LIVE TO 120?

As mentioned previously, scientists believe that about 25 percent of an organism's longevity potential is determined by genetics. Children of long-lived parents are likely themselves to reach advanced ages. Siblings of centenarians have much higher lifespans compared with the lifespans of the general population. In addition, children of centenarians have fewer health problems than children of people who die in their 70s. One study showed that the incidence of high blood pressure and heart disease was 26 percent and 13 percent, respectively, in children of centenarians, but 52 percent and 27 percent in children of shorter-lived parents. Although these observations suggest a familial component to attaining extreme old age, they do not differentiate between genetic causes and environmental ones.

Longevity advice given by centenarians is often unenlightening. The world's oldest documented human – Jeanne Calment, who died in 1997 at the age of 122 – smoked and ate two pounds of chocolate per week until the age of 119 (See Box: Tale of a Supercentenarian). Many other centenarians ignore all the dietary and behavioral guidelines meant to extend our healthy lifespans, but still live to 100. A study of centenarians showed that 30 percent were overweight, none exercised regularly, and none were vegetarians or ate yogurt. Studies such as this suggest that there is something about the genetic makeup of healthy centenarians that allows them to bypass the requirements for sensible lifestyle and still defy aging.

Over 90 percent of centenarians are women. Men have higher death rates than women at all ages. Part of this may be due to genetic differences between males and females, but a majority of it appears to be the result of behavioral differences. Between the ages of 14 and 25, the male death rate exceeds that of females three-fold. This is due to higher rates of vehicle accidents, murders and deaths that arise from violent hobbies and risky occupations. Until menopause, women have much lower death rates due to heart disease. More men than women smoke and suffer from alcoholism, both of which increase the rates of heart disease, liver disease and lung cancer. Women tend to have healthier diets and seek medical attention when needed. In addition, men are less likely to enjoy close social interactions and accept support from peers and family. Supportive social interactions have been shown to reduce feelings of isolation and contribute to improved mental and physical health.

It is perhaps obvious that economic prosperity contributes to a long and healthy life. The advantages of living in wealthy, developed nations are many: widespread sanitation, availability of antibiotics, healthcare, education and adequate nutrition. Wealthy nations such as Japan and Australia have the highest average and maximum lifespans; third world nations such as Sierra Leone and Malawi have the lowest. The link between economic factors and longevity was seen in Russia after the fall of communism, when the life expectancies of men dropped from about 65 to 58, between 1990 and 1995. In the United States, certain socioeconomic groups such as rural African Americans, Native Americans and inner city poor have much lower life expectancies than the mainstream caucasian population.

Last, but not least, chance is a strong determinant in exceeding the age of 100. The longer one lives, the more chances there are to contract a fatal infection, accumulate mutations that lead to cancer or die by accident. Centenarians are not only healthier than average, but they are also unusually lucky.

Although most of the longevity factors discussed above are beyond an individual's control, there are still a few strategies that one can use to increase the odds of having a long, healthy life – maintain a healthy weight, exercise regularly and avoid smoking.

Questions About Aging and Society

WHAT ARE THE SOCIAL, ECONOMIC AND POLITICAL IMPLICATIONS OF EXTENDING THE HUMAN LIFESPAN?

At present, we can draw two general conclusions about scientifically based antiaging interventions. First, no treatment, drug or lifestyle has been shown to stop or arrest the aging process and extend the maximum lifespan of human beings. Second, research on the physiology and genetics of aging has led to interventions that retard aging and extend the lifespans of laboratory animals. Scientists are optimistic that one or more human antiaging therapies will emerge within the next 10 to 20 years. If these therapies resemble current interventions that are successful in experimental animal models, they may extend average and maximum human lifespans from 30 to 70 percent. This translates into an increase in average lifespan, in developed countries, of up to 130 years and a maximum lifespan of up to 200 years. Some optimistic researchers predict that science may find ways to extend healthy human lives into the hundreds of years. Because these therapies will be attacking the basic mechanisms of aging, they will likely result in longer disease-free years of life prior to the onset of age-related conditions such as dementias and cardiovascular disease. Laboratory animals whose lives are extended with genetic manipulation or calorie restriction do not suffer longer periods of old age, but they do have longer disease-free lives. The future potential of longer healthier lives is also supported by observations of centenarians, who are usually healthy and active through their 90s and 100s. These unusually long-lived individuals tend to have only short periods of ill health just prior to death. If future antiaging therapies allow more people to live long, healthy lives similar to those of centenarians, these longer-lived people may not be any more of a drain on the health care system than those who are presently 80 years old.

This does not necessarily mean that future antiaging therapies will be devoid of side effects. The price we pay for longer lives may be a loss of fertility or lower metabolic rates (See Box: Calorie Restriction and Longevity); however, it is also possible that remedies for these side effects will be developed, or that people will tolerate them in exchange for longevity.

In addition, it seems likely that the extension of the human lifespan will have effects upon every aspect of society. One concern is the effect of longer lives on the world's population. Some foresee a day when population growth will be exacerbated by growth in the elderly segment of society. Those who argue that this will not be a problem state that affluent countries are not suffering population growth and that birth rates are lower than replacement levels. An increase in the number of elderly people will not cause an overpopulation problem within those societies. A different scenario could exist in countries in which overpopulation is currently a problem. However, since these countries tend to be less affluent than those that currently have low birth rates, and since antiaging therapies may be expensive and not available to everyone, the effects of antiaging therapies on these populations may not be great. This potential division between rich and poor will likely create problems, as discussed in the next section.

Another concern is the effect of longer lives on the US social security system. Funding for pension plans is now based on a limited number of years of retirement and a large enough working population to support these pensions. In the future, our present pattern of employment for 30 to 40 years followed by retirement with adequate pension may be impossible to sustain. People may have to either work for most of their lives or undertake personal retirement strategies that would allow them to retreat from employment after about 50 years. Retirement may not be a permanent situation for long-lived people. It may be more of a sabbatical period, during which they develop new skills and education with which to begin a new career.

Potentially positive economic outcomes could follow from the presence of a large population with longer, healthier lives. People with special skills and accumulated knowledge could continue to contribute to society for decades longer than people do today. People might be more willing to enter long educational periods in order to undertake highly skilled professions, increasing the amount of experience and talent in society. Longer lives could also cause a shift in the nature of work, as people would have time to train for and develop several careers within their lifetimes.

Other effects of extended longevity would be changes in family structures. Families would likely consist of multiple generations with different dynamics of interactions and inheritance. This might tip the balance of power and wealth to the longest-lived, and hence the longest-producing, members of society. Our current educational structures would also change, as people would have more time to explore more fields and train for new careers.

Some predict that the emergence of effective antiaging therapies could potentially cause social upheavals. The demand for these therapies would likely be universal and great. Unless sufficient drugs or procedures were available simultaneously for all who elected to extend their lives, a new "have - have not" division in society would emerge. People would fight to become part of the long-lived group, rather than part of the group destined to age and die young. There could also be conflict about the world-wide distribution of antiaging therapies, regardless of cost. The current tragic inequality in the availability of affordable AIDS drugs to underdeveloped nations could foreshadow a similar scenario in regards to antiaging drugs.

Another potentially negative effect of life extension might be an increase in social conservatism. Cultural and political changes often require the influence of a new generation with new ideas. If younger people cannot seize power from the majority of older people, and those with extended lives tend towards social and political stability as they do now, social change may be retarded.

Of course, many of the future economic, political and social effects of antiaging therapies will depend on the degree of the lifespan extension. A 10 to 20 percent increase in lifespan may not have serious social consequences; however, therapies that prevent or even reverse aging would likely have immediate and long-term effects on society. It may be useful to consider these scenarios now, before science presents us with the first truly effective aging intervention.

WHAT ARE THE ETHICS OF EXTENDING HUMAN LIFESPAN?

No one wants to age and die. As Jonathan Swift reminded us, "Every man desires to live long; but no man would be old." This desire for the Fountain of Youth encourages people to spend billions of dollars each year to purchase useless, and sometimes harmful, elixirs that are supposed to cure aging or restore youthful appearance. Paradoxically, though, many individuals oppose the idea that we should cure aging. When presented with a theoretical prospect of immortality, a long list of sobering objections arise.

One interesting objection to extreme life extension is the boredom argument. This argument states that people who live to be hundreds of years old would simply become bored and their lives would become devoid of meaning. After a while, they would experience everything they wished to experience, after which their lives would stagnate. In this argument, death is a stimulus for an intense and meaningful life. Longevity proponents dismiss this argument by saying that extended lifespans would not be obligatory for those who did not wish to extend their lives. In addition, they believe that youthful, unaging brains would continue to grow and develop, rather than become rigid and conservative.

If longevity interventions involved regenerative medicine – replacement of body parts with artificial organs or stem cell transplants – would this compromise the concept of "self?" How much could one's body be replaced, including parts of the brain, before a person's individuality would be compromised? In a world with no aging, an ever-youthful human brain would presumably continue to change with new knowledge and experiences. Would individuals change so dramatically over several hundred years that they would cease to be "themselves?" If so, would this be sufficient reason to refuse antiaging therapy?

Another argument against life extension is that therapies would be so expensive that only a portion of the world's population could afford them, creating another global injustice – this time between the mortals and immortals. Some argue that we should not engage in activities so profound as longevity extension unless we can offer the benefit equally to all. Although a principled moral argument, similar arguments have not made much inroad into public policies regarding universal availability of advanced medical treatments such as heart transplantation or modern HIV/AIDS drugs. Whether this argument would have weight in the adoption of antiaging therapies is questionable.

If life extension interventions were available, should society make them available to everyone, including political tyrants and criminals guilty of murder or genocide? If life extension interventions were freely available to all people, how would society treat those who refused therapy? Would this be a form of suicide and be denounced by religion and the law?

Would efficient life extension methods mean an end to reproduction for most people? If humans were able to live to hundreds, maybe thousands, of years, would it be reckless to reproduce at our current rates, leading to overpopulation? How would we decide who would be allowed to reproduce, when and how often?

Another ethical question is whether the desire to extend our lives is selfish and unworthy. Humans often value self-sacrifice and martyrdom for noble causes. Would this mean that those who chose to extend their lives, perhaps combining life extension techniques with avoidance of risky situations, be looked down upon by society as narcissistic? Would society consider these people to be selfishly consuming more than their share of limited world resources in order to support their extremely long lives? Along a similar line, some question whether life extension research is a waste of money, preferring that we put resources into developing health improvements for those who are already elderly. Others believe that we should put more money into improving the lives of those whose economic resources are limited – in both developed and developing nations. In this view, antiaging therapies have the same problems as other expensive medical breakthroughs that are only available to the wealthy.

Scientists who have considered many of these arguments feel that the desire for life extension is so strong that research into aging and life extension is unstoppable. People will do whatever is necessary to attain a long life, regardless of the social and ethical consequences. Although it is prudent to consider ethical questions about life extension, we may have little control over the course of scientific advancements or the nature of their implementation.

HOW DO CULTURES DIFFER IN THEIR ATTITUDES TOWARDS AGING AND THE AGED?

There are many reasons that most of us fear aging. The most obvious are that we do not want to die and that aging makes us face our mortality. In addition, we fear aging because it can bring disability, pain and social stigmatization. Although the incidence of many diseases such as cancer and heart disease increase with age, it is perhaps the social aspects of aging that are the most difficult for older people to deal with.

In western, industrialized societies, age is not held in high esteem. Surveys show that very few children in western industrial countries know older people outside their family group. This limited knowledge is accompanied by negative attitudes towards the elderly. In addition, negative attitudes towards age extend into the older population itself. A Harris Poll found that more than 90 percent of Americans over the age of 65 objected to the use of the term "old." Less than half found the terms "senior citizen" and "retired person" acceptable. About the time of the industrial revolution, new language emerged to describe the elderly – codger, gaffer, old fogy and geezer. This language development appears to have accompanied a general cultural shift from respect for the elderly to disdain.

At present, the image of the elderly in American society is still generally negative. Older people are perceived as being less healthy, less alert, lonely, impoverished, bored and unemployed. Interestingly, older people (defined as those over 65) seem to accept these stereotypes as well – at least for others of their age group. In a 1994 survey conducted by the American Association of Retired Persons, more than half of older individuals surveyed agreed that older people suffered from poor health, loneliness and poverty, but fewer than 15% agreed that these traits applied to themselves. It has been noted that the mass media under-represents the elderly. For example, about 3 to 5 percent of television show characters are over the age of 70; whereas this group represents about 10 percent of the population. There is no "rite of passage" (a ritual that grants membership into a socially accepted group) for elderly people. The elderly are marginal in our society. More recently, the negative stereotype of older people as helpless, bored and sick has been augmented by a new stereotype: older persons are greedy and selfish, enjoying carefree, luxurious lives at the expense of the younger generations.

The roots of these negative images appear to be related to our culture's emphasis on individuality, competitiveness, productivity and the propensity to spend on consumer products. These qualities are associated with youthful people, not the elderly. Some analysts believe that urbanization and industrial mass production are the basis for marginalization of the elderly. These forces tend to disrupt economic dependency and extended families. The older people that are respected in our society are those who continue to perform in the realms usually associated with youth – those who continue to have productive work and professional lives, those who prominently serve the community, those who continue to strive for personal growth and those who partake in adventurous activities.

Negative images and stereotypes of aging people are more than interesting sidelights: these attitudes also alter the behavior of the aged. It is well known that a person's self image, and hence behavior, is partly determined by how that person is treated. Furthermore, the way we treat people comes from how we perceive them. All through our lives, we define ourselves by our age group. Educational timetables (kindergarten, then school and college), marriage, parenting, career and retirement are all linked to age. Attitudes towards aging also affect political decisions about how society cares for older people and how it constructs medical care systems. Some studies suggest that positive attitudes may affect age-related abilities such as memory. Elderly Chinese people who have positive attitudes towards aging perform as well as younger Chinese people on memory tests. In contrast, elderly Americans who have negative attitudes towards aging perform poorly on these tests compared with younger Americans. Negative stereotypes may also affect longevity. Some studies find a correlation between having a positive attitude towards aging and a person's lifespan.

To define society's attitudes on aging, it is useful to consider the attitudes towards aging in cultures that do not emphasize individuality, but rather interdependence. In traditional Japanese culture, older people are considered to have wisdom, are granted respect and are allowed to pursue creative enterprises. Grandparents play roles in childrearing and families are expected to respect, care for and obey their aging relatives. As a result, most elderly Japanese people are cared for in the homes of their children. Although care facilities for elderly people are still rare in Japan, more have appeared in recent years.

Although certainly diverse, Native American cultures tend to hold the elderly in high esteem. Older people are believed to have wisdom and knowledge that can benefit the family and group. In some Native American cultures, grandmothers are the centers of the family and help to raise the grandchildren.

The Chinese also venerate their elderly citizens. Older people are seen as wise and their advice is sought in decision making. Traditional Chinese attitudes of obedience, reverence for family and social harmony directly affect the treatment of the elderly. Aging parents are usually cared for by their families and this is a source of pride for the family. These attitudes are reflected in political policies. The Chinese Marriage Law states that children are required to support their parents. In recent years, the government has begun to offer public support for older people by providing basic income and housing for people who are unable to rely on family support. The Chinese trend towards smaller families with fewer children, however, may put a strain on traditional approaches to caring for the elderly.

Not all cultures, however, revere the elderly. Although the majority of African societies consider old age to be an asset, some view it as a sign of misfortune. In a few cases, leaders who became old and weak were sacrificed so that they could be succeeded by younger leaders. In addition, the general attitudes within cultures such as the Japanese, although positive towards aging, sometimes mask a latent hostility. Elder abuse occurs in these cultures, perhaps due to the heavy responsibilities imposed by the expectation that the elderly should be cared for by their children.

ARE THERE ADVANTAGES TO AGING?

The negative stereotypes of elderly people that pervade our culture are set to change. Within the next 20 years, the number of people over the age of 65 will double. The demographic and economic force of the baby boomers entering old age is expected to cause major shifts in society's attitudes about age. In addition, research on aging is revealing some surprising data that challenge our conceptions of the elderly.

One stereotype of the elderly is that they are in poor health. However, recent studies show that the rates of disability for older people are in the range of 5 percent for those between 60 and 65, about 10 percent for those between 70 and 75 and about 20 percent for those over the age of 85. Therefore, most older people maintain sufficient physical and mental function to live independently for most of their later years (Figure 9). It is documented that people lose about

Figure 9
The majority of older people have sufficient mental and physical fitness to live productive, independent lives.
Source: Getty Images, Time & Life Pictures

1 to 2 percent of their muscle strength per year after the age of about 40. However, these losses can be reduced by simple exercise and strength training.

Another stereotype is that older people are mentally impaired, unable to perform well in memory and reaction tests and unable to learn "new tricks." Although there is a documented decline in mental performance with age, it may not be as pronounced as once believed. Recent data show that the brains of older people do not lose as many neurons as once thought and that new neurons can appear in older brains. In addition, the nature of the laboratory tests that measure mental abilities can negatively alter older people's performances. For example, most psychological tests are conducted late in the day, which is optimum for younger people. Most older people, however, function better in the mornings. The differences in memory scores between young and old people are reduced by one-half when the tests are conducted in the morning. In addition, older people are influenced by aging stereotypes. After reading material that praises the positive aspects of aging, older people perform up to 30 percent better in memory tests than after reading material that reinforces the negative aspects of aging. Older individuals are also influenced by the nature of the material that comprises a memory test. When asked to remember details about stories that engage their interest – such as stories about finding retirement homes – older people perform as well as younger ones. They also perform well on tasks that they practice regularly, such as playing a musical instrument or giving a lecture (Figure 10). Older people appear to process information differently than younger people. They are more likely to remember images or details that are linked to positive emotions and less likely to remember

Figure 10
Older people perform well when using skills they practice regularly. Source: Getty Images, Taxi

negative images or events. Studies indicate that adequate physical, social and mental activity helps protect older people from the cognitive declines that often accompany aging.

Older people perform better than younger people in a number of areas and the skills they have improve with age. Older people appear to have greater abilities to judge character and greater skills in interpersonal relationships. They are skilled at analyzing complicated social situations and devising solutions. They also have better verbal abilities than younger people. Despite the notion that older people are usually depressed and lonely, studies show that the elderly are more often in good mental health and are more optimistic and happier than young people. Older people tend to live for the moment and seek pleasure in the simpler things of life.

The idea that extreme old age (90 years and over) is inevitably characterized by physical and mental deterioration is also being challenged. Studies of centenarians reveal some surprising facts. Most centenarians are healthy and active throughout their lives. Many are active sexually, have employment and partake in the arts or outdoor activities. In many cases, centenarians are healthier than many people in their late 70s and 80s. One study showed that the rate of Alzheimer's disease in centenarians is only about 30 percent – lower than the rates in younger elderly people. Centenarians usually have a short period of disability just prior to their deaths. The causes of their deaths are often acute infectious diseases such as pneumonia, rather than long-term disabilities such as cardiovascular disease. Although one could argue that centenarians represent an unusually robust segment of the population that progresses to extreme old age because of their unusual mental and physical strength, they also illustrate that old age does not necessarily have to be plagued with infirmity. If scientifically based antiaging therapies are developed, and these retard or reverse the basic causes of biological aging, more people may become like present-day centenarians – resistant to the cellular damage that makes them susceptible to age-related diseases.

In the last decade we have seen the beginning of changes in how society views aging and the elderly. This will undoubtedly accelerate as science comes closer to understanding the true causes of aging and develops strategies to retard or reverse the aging process.

References and Resources

Books

Benecke, M. The Dream of Eternal Life. Columbia University Press, New York, 2002.

Clark, W.R. A Means to an End: The Biological Basis of Aging and Death. Oxford University Press, New York, 2002.

Guarente, L. Ageless Quest: One Scientist's Search for Genes that Prolong Youth. Cold Spring Harbor Laboratory Press, Cold Spring Harbor, New York, 2003.

Hayflick, L. How and Why We Age. Ballantine Books, New York. 1994.

Masoro, E.J. and Austad, S.N., editors. Handbook of the Biology of Aging, 5th edition. Academic Press, San Diego, CA. 2001.

Olshansky, S.J. The Quest for Immortality: Science at the Frontiers of Aging. Norton, New York, 2001.

Publications

Abbott, A. 2004. Growing old gracefully. *Nature* **428**:116–118.

Beckman, K.B. and Ames, B.N. 1998. The free radical theory of aging matures. *Physiol. Rev.* **78**: 547–581.

Bird. J. et al. 2003. Can we say that senescent cells cause aging? *Experimental Gerontology* **38**:1319–1326.

Campisi, J. 2003. Cancer and aging: rival demons? *Nature Reviews Cancer* **3**:339–349.

De Grey, A., editor. 2004. Strategies for engineered negligible senescence: why genuine control of aging may be foreseeable. *Ann. N.Y. Acad. Sci.* (Review articles) **1019**:1–592.

Dumble, M. et al. 2004. Insights in aging obtained from p53 mutant mouse models. *Ann. N.Y. Acad. Sci.* **1019**:171–177.

Finkel, T. and Holbrook, N.J. 2000. Oxidants, oxidative stress and the biology of aging. *Nature* **408**:239–247.

Finkel, T. 2003. A toast to long life. *Nature* **425**:132–133.

Guarente, L. 1999. Mutant mice live longer. *Nature* **402**:243–245.

Guarente, L. and Kenyon, C. 2000. Genetic pathways that regulate aging in model organisms. *Nature* **408**:255–262.

Harris, J. 2004. Immortal ethics. *Ann. N.Y. Acad. Sci.* **1019**:527–534.

Hasty, P. et al. 2003. Aging and genome maintenance: lessons from the mouse. *Science* **299**:1355–1359.

Hayflick, L. 2000. The future of aging. *Nature* **408**:267–269.

Hayflick, L. 2003. Living forever and dying in the attempt. *Experimental Gerontology* **38**:1231–1241.

Hekimi, S. and Guarente, L. 2003. Genetics and the specificity of the aging process. *Science* **299**:1351–1354.

Helmuth, L. 2003. The wisdom of the wizened. *Science* **299**:1300–1302.

Itahana, K. et al. 2004. Mechanisms of cellular senescence in human and mouse cells. *Biogerontology* **5**:1–10.

Kipling, D. et al. 2004. What can progeroid syndromes tell us about human aging? *Science* **305**:1426–1431.

Kirkwood, T.B.L. and Austad, S.N. 2000. Why do we age? *Nature* **408**:233–238.

Lane, M.A. et al. June 2004 special edition. The serious search for an antiaging pill. *Scientific American* **14**(3):36–41.

Longo, V.D. and Finch, C.E. 2003. Evolutionary medicine: from dwarf model systems to healthy centenarians? *Science* **299**:1342–1345.

P. D. June 2004 special edition. The cryonics gamble. *Scientific American* **14** (3):82–84.

Perls, T.T, et al. 1997. Middle-aged mothers live longer. *Nature* **389**:133.

Perls, T.T. June 2004 special edition. The oldest old. *Scientific American* **14**(3): 6–11.

Sharpless, N.E. and DePinho, R.A. 2004. Telomeres, stem cells, senescence, and cancer. *J. Clin. Invest.* **113**:160–168.

Tatar, M. et al. 2003. The endocrine regulation of aging by insulin-like signals. *Science* **299**: 1346–1351.

Wright, K. 2003. Staying alive. *Discover* **24**(11): 64–71

Web Sites

http://www.benbest.com/lifeext/aging.html
Mechanisms of aging, by Ben Best

http://www.stlcc.cc.mo.us/mc/users/vritts.aging.html
Culture and aging, by V. Ritts

http://www.who.int/inf-pr-2000/en/pr2000-life.html
World Health Organization issues new healthy life expectancy rankings.

http://www.methuselahfoundation.org/
The Methuselah Foundation, originator of the Methuselah Mouse Prize

http://aeiveos.com/longevity/jlcinfo.html
Information on Jeanne Calment, the world's oldest documented centenarian

http://www.pbs.org/stealingtime
PBS television series dealing with the science of human aging

http://www.walford.com/index.html
Information about calorie restriction and calorie restriction proponent, Roy Walford

http://www.calorierestriction.org
Details on the practice of calorie restriction

http://www.longevitymeme.org
News articles on longevity and aging

http://www.immortality.org
News and links to sites dealing with aging and life extension

http://www.sciam.com/agingstatement.cfm
The Truth about Human Aging: a position statement by 51 scientists on aging and antiaging remedies

http://www.aleph.se/Trans/Individual/Life/
Links on immortality and life extension

http://sageke.sciencemag.org
Online resources for the science of aging, run by the journal *Science* and the American Association for the Advancement of Science.

Photo Credits